SpringerBriefs in Applied Sciences and Technology

SpringerBriefs present concise summaries of cutting-edge research and practical applications across a wide spectrum of fields. Featuring compact volumes of 50 to 125 pages, the series covers a range of content from professional to academic.

Typical publications can be:

- A timely report of state-of-the art methods
- An introduction to or a manual for the application of mathematical or computer techniques
- A bridge between new research results, as published in journal articles
- A snapshot of a hot or emerging topic
- An in-depth case study
- A presentation of core concepts that students must understand in order to make independent contributions

SpringerBriefs are characterized by fast, global electronic dissemination, standard publishing contracts, standardized manuscript preparation and formatting guidelines, and expedited production schedules.

On the one hand, **SpringerBriefs in Applied Sciences and Technology** are devoted to the publication of fundamentals and applications within the different classical engineering disciplines as well as in interdisciplinary fields that recently emerged between these areas. On the other hand, as the boundary separating fundamental research and applied technology is more and more dissolving, this series is particularly open to trans-disciplinary topics between fundamental science and engineering.

Indexed by EI-Compendex, SCOPUS and Springerlink.

Youngsub Lim

Alternative Fuels for Environmentally-Friendly Ships

Hydrogen, Ammonia, Bio-fuels and E-fuels

 Springer

Youngsub Lim
Department of Naval Architecture
and Ocean Engineering
Seoul National University
Seoul, Korea (Republic of)

ISSN 2191-530X   ISSN 2191-5318 (electronic)
SpringerBriefs in Applied Sciences and Technology
ISBN 978-3-031-85081-3   ISBN 978-3-031-85082-0 (eBook)
https://doi.org/10.1007/978-3-031-85082-0

This Springer imprint is published by the registered company Springer Nature Switzerland AG
The registered company address is: Gewerbestrasse 11, 6330 Cham, Switzerland

If disposing of this product, please recycle the paper.

Preface

The shipping and shipbuilding industries are facing an unprecedented period of change. To regulate greenhouse gas (GHG) emissions, the energy efficiency design index (EEDI) has been applied to newbuild ships from 2013 and the energy efficiency existing ship index (EEXI) and carbon intensity index (CII) also have been applied to existing ships from 2023. Now ships that do not comply with the GHG regulations cannot operate.

Furthermore, the International Maritime Organization (IMO) adapted the "2023 IMO Greenhouse Gas Strategy" at the 80th Environmental Protection Committee (MEPC) in 2023, which significantly strengthens the GHG reduction goal to net-zero by or around 2050. In addition, in recognition of the problems in assessing the GHG emission intensity of the conventional Tank-to-Wake (TtW) process, the introduction of a new Well-to-Wake (WtW) GHG emissions intensity based on life-cycle assessment (LCA) is being considered, which has led to a new phase of GHG reduction for ships. This is a shift from the traditional focus on "what fuel is used" to a new paradigm of "how it's produced and used." This decision is also leading to a complete rethink of the traditional fossil fuel-based shipping strategy, and a number of controversies surrounding alternative marine fuels and related technologies have emerged.

Hydrogen is a prominent decarbonized alternative fuel, but most of the hydrogen currently available is gray hydrogen, which is produced by reforming natural gas and therefore has a higher WtW GHG emissions intensity than even fossil fuels. Green hydrogen is not yet economically feasible, and it is not well-understood that liquefaction and transport of hydrogen requires additional energy. Ammonia, which is more economical to transport and use, has been proposed as an alternative solution, but the gray ammonia also has a higher WtW GHG emissions intensity than fossil fuels. Moreover, additional technical considerations, such as toxicity issues and the generation of additional GHGs from nitrous oxide, are rarely mentioned. Biofuels and e-fuels are also considered as alternative fuels, but it is not well-known that quantitative sustainability criteria are currently being proposed and that in the future fuels that do not meet the criteria may not be approved as biofuels or e-fuels, even if they are produced from biomass or green hydrogen. While there is growing interest in carbon capture, utilization and storage (CCUS) and onboard carbon capture and storage (OCCS) technologies, there are also limitations.

The knowledge required for alternative fuels for environmentally friendly ships is diverse and requires an interdisciplinary understanding. I have tried to summarize in this book the changing international situation and various knowledge, for those interested in environmentally friendly ships and alternative fuels. I hope it will be useful to all those involved in the shipping and shipbuilding industries.

Finally, I would like to thank my lovely wife, Sujin.

Seoul, Korea (Republic of)   Youngsub Lim

Competing Interests The author has no competing interests to declare that are relevant to the content of this manuscript.

Contents

Chapter 1
Alternative Fuels and 2023 IMO GHG Strategy

Abstract The definition of alternative fuels is evolving from a material-based approach (focusing on what is used) to a process-based approach (emphasizing how fuels are produced). For alternative fuels to be considered suitable for environmentally friendly ships in future, it seems that they need to meet three following criteria. First, to truly estimate the impact on climate change, carbon dioxide (CO_2) equivalent greenhouse gas (GHG) emissions considering global warming potential should be evaluated, rather than just CO_2 emissions from fuel use. Second, well-to-wake GHG emissions should be evaluated considering lifecycle assessment from the production to use of a fuel. Third, specific quantitative criteria will be required to qualify a fuel as an alternative fuel, ensuring meaningful reductions in GHG emissions. The recently announced 2023 IMO GHG strategy, along with discussions at the Marine Environment Protection Committee shows these features well. This book introduces alternative fuels for environmentally friendly ships including the topics in liquefied natural gas, hydrogen, ammonia, biofuels, e-fuels and carbon capture, utilization, and storage.

Keywords Alternative fuels · Global warming potential · Life-cycle assessment · FuelEU maritime · 2023 IMO GHG strategy · Well-to-wake GHG emissions

1.1 What Are Alternative Fuels?

Recently, the term "alternative fuels" has been used frequently in various media. At the same time, a lot of controversies and contradictory statements also appear. One day, you might read an article that says biofuels are an alternative fuel of the future, and the next day you might read a criticism that says biofuels are not environmentally friendly. Why does this controversy persist? It's because people have been using different definitions of alternative fuels. The term alternative fuel literally means "a new fuel that can replace an existing fuel," which is a vague definition that varies from person to person, and as a result, the fuels it refers to have changed over time.

© The Author(s), under exclusive license to Springer Nature Switzerland AG 2025
Y. Lim, *Alternative Fuels for Environmentally-Friendly Ships*,
SpringerBriefs in Applied Sciences and Technology,
https://doi.org/10.1007/978-3-031-85082-0_1

In 2014, the EU European Parliament defined alternative fuels as follows. Probably this is the definition that best aligns with what most people think of as alternative fuels.

> Alternative fuels are defined as fuels or power sources which serve, at least partly, as a substitute for fossil oil sources in the energy supply to transport and which have the potential to contribute to its decarbonisation and enhance the environmental performance of the transport sector

The definition also gave some examples as follows:

- Electricity
- Hydrogen
- Biofuels
- Compressed Natural Gas (CNG)
- Liquefied Natural Gas (LNG)
- Liquefied Petroleum Gas (LPG).

This definition is based on the idea that "what is used" as a fuel determines whether or not it is an environmentally friendly alternative fuel. It's a fundamental source of the confusion that has led to so much of the debate.

For a simple example, let's think about electricity from various sources, as shown in Fig. 1.1. When people think of electricity as an alternative fuel, they think of electricity generated from renewable sources like wind or solar power. Can the electricity generated by a coal-fired power plant be a green alternative fuel? Most people would say no. In other words, it doesn't matter whether it's "electricity" or not when we think of alternative fuels. What matters is "how it's produced and used."

The situation is similar for other alternative fuels. The conventional substance-based definition of alternative fuels cannot fundamentally avoid the confusion because it does not take into account the production process. After years of debate, the criteria for alternative fuels is now shifting from "what's used" to "how it's produced and used." In other words, the way we define alternative fuels is shifting from what we use to how they are made. To reflect this perspective, we can refine the definition of alternative fuels a bit further and say something like this.

Fig. 1.1 Electricity from various energy sources

> **Alternative fuels are defined as fuels (or power sources) which contribute to decarbonization by emitting significantly less greenhouse gases (GHGs) per unit of energy provided than fossil fuels, including their supply chain from production to use.**

Here, we can find three important issues. First, what substances should be included in the assessment of GHG emissions? Only carbon dioxide? Second, how the supply chain can be defined from where to where? Third, if a fuel can be an alternative fuel only if it reduces GHG emissions significantly less than fossil fuels, how much does it have to reduce? Could a 5% GHG reduction qualify as an alternative fuel? If not, how about 10%? 20%? These questions have been at the center of debate over the past decade. In this chapter, we will briefly summarize the answers to these key questions and summarize why this perspective is necessary to understand alternative fuels for environmentally friendly ships.

1.2 Global Warming Potential (GWP)

The first issue is the scope of GHGs, which was defined relatively early on. The Kyoto Protocol, adopted in 1997 under the United Nations Framework Convention on Climate Change (UNFCCC) (see Chap. 2 for more details on the Kyoto Protocol), identified the following six major GHGs as representative anthropogenic GHGs, whose atmospheric concentrations have increased as a result of human activities and thus contribute significantly to the greenhouse effect.

- Carbon dioxide (CO_2)
- Methane (CH_4)
- Nitrous oxide (N_2O)
- Hydrofluorocarbons (HFCs)
- Perfluorocarbons (PFCs)
- Sulfur hexafluoride (SF_6).

Of course, there are many other GHGs besides these. Water, for example, is known as the greenhouse gas with the greatest impact on the greenhouse effect because of the large amounts of water present in the atmosphere. However, water is not considered an anthropogenic greenhouse gas because it evaporates or condenses easily in response to global climate conditions, rather than human activities.

The greenhouse effect caused by each GHG is not identical. To compare the relative contribution of a GHG to the greenhouse effect, an index of Global Warming Potential (GWP) has been developed. GWP is defined by a mathematical formula based on radiative forcing, but if we take the math out of it for fast understanding, the meaning of GWP is defined as follows.

The **GWP is defined as an index measuring the effect to global warming following an emission of a unit mass of a given substance, accumulated over a chosen time horizon, relative to that of the reference substance, carbon dioxide (CO_2)**

$$\text{GWP}[t] \text{ of } x = \frac{\text{Global warming effect of 1 g of substance } x \text{ over a period of } t}{\text{Global warming effect of 1 g of reference substance, } CO_2}.$$

(1.1)

The GWP value can be estimated over different time scales, such as 20, 100, and 500 years. The 100-year GWP scale is now used as the standard in most international policies. Table 1.1 shows the GWP of various materials as reported in IPCC 6th Assessment Report. The GWP100 of methane is almost 30 times greater than that of CO_2. Substances such as HFCs, PFCs, and CFCs are man-made substances, and have GWP thousands of times greater than that of CO_2. Because different materials have very different impact on the greenhouse effect as describe, GWP is a useful index.

GWP makes it easy to convert the greenhouse gas emissions of complex mixtures into CO_2 equivalent GHG emissions. For example, exhaust gas from a certain fossil fuel contains 100 g of CO_2 and 1 g of methane, it can be converted to CO_2 equivalent GHG emissions of 129.8 g_{CO_2eq}, which is easy to understand.

$$CO_2\text{eq GHG emissions} = \sum \text{GWP}_i \cdot m_i = 1 \times 100 + 29.8 \times 1 = 129.8 \ g_{CO_2eq}$$

However, it is important to note that even for the same substance, GWP values may be adjusted depending on the reference as shown in Table 1.2, as new opinions become available at the time of the assessment. The current scientific debate on greenhouse gases is still ongoing, so it's natural for different researchers or institutions to report different values. This book cites the most recently published IPCC 6th Assessment

Table 1.1 GWP of representative greenhouse gases [4]

Name of compound	Molecular formula	Lifetime (year) in atmosphere	GWP20	GWP100	GWP500
Carbon dioxide	CO_2		1	1	1
Methane (fossil origin)	CH_4	11.8	82.5	29.8	10
Methane (non-fossil origin)			80.8	27.2	7.3
Nitrous oxide	N_2O	109	273	273	130
HFC-134a	$C_2H_2F_4$	14	4,144	1,526	436
CFC-11	CCl_3F	52	8,321	6,226	2,093
PFC-14	CF_4	50,000	5,301	7,380	10,587

Table 1.2 Different GWP100 on IPCC reports

Name of compound	GWP100		
	IPCC 4th report [2]	IPCC 5th report [3]	IPCC 6th report [4]
CO_2	1	1	1
CH_4	25	28, 34	29.8 (fossil origin)
			27.2 (non-fossil origin)
N_2O	298	265, 298	273

Report [4], but figures from the 5th or 4th Assessment Reports are often still used, depending on the source.

1.3 Life-Cycle Assessment (LCA)

The second issue is the scope of the assessment of GHG emissions, from where to where. If different institutes or different products use different scopes to assess GHG emissions, the same product would end up with different GHG emissions. To avoid this confusion, many institutions generally follow the scope defined by the Greenhouse Gas Protocol (GHG Protocol).

The GHG Protocol is a standard for measuring GHG emissions from an organization. It was established by the World Resources Institute (WRI) and the World Business Council for Sustainable Development (WBCSD) in 1990 in response to demand for the need of an international standard for carbon accounting.

The GHG Protocol classifies the GHG emissions into three scopes, as shown in Table 1.3 and Fig. 1.2. Scope 1 emissions are direct emissions from resources owned or operated by an organization that are emitted directly into the atmosphere as a result of the organization's activities. Scope 2 emissions are indirect emissions from the generation of utilities that are purchased by the organization: use of utilities such as electricity, steam, heat, and cooling that are purchased by the organization. Scope 3 emissions are GHG emissions from the entire supply chain of the organization's value chain, from the supply of raw materials to distribution, storage, and even disposal of the product.

Table 1.3 Scope 1, 2, and 3 defined in GHG Protocol [1]

Scope	Definition
Scope 1	Direct emissions: GHG emissions from resources owned or operated by an organization
Scope 2	Indirect emissions: GHG emissions from the generation of energy such as electricity, steam, heat, and cooling that an organization purchases to use
Scope 3	Supply chain emissions: GHG emissions from the organization's entire supply chain, including the supply of raw materials, distribution, storage, and disposal of products

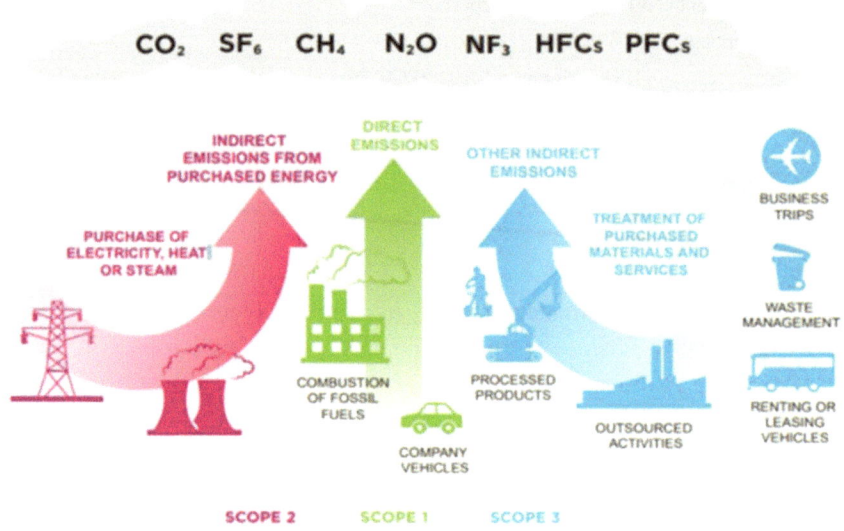

Fig. 1.2 Scope 1, 2, and 3 defined in GHG Protocol [5]

In the past, Scope 1 and 2 GHG emissions have been the main focus of discussion and estimation. However, with research showing that Scope 3 emissions could account for more than 70% of total GHG emissions as shown in Fig. 1.3, there is a growing consensus of the assessment and reduction of Scope 3 emissions. For example, global oil companies such as British Petroleum (BP), Equinor, Total, and Shell, are announcing scope 3 emissions reduction targets for 2020 and beyond.

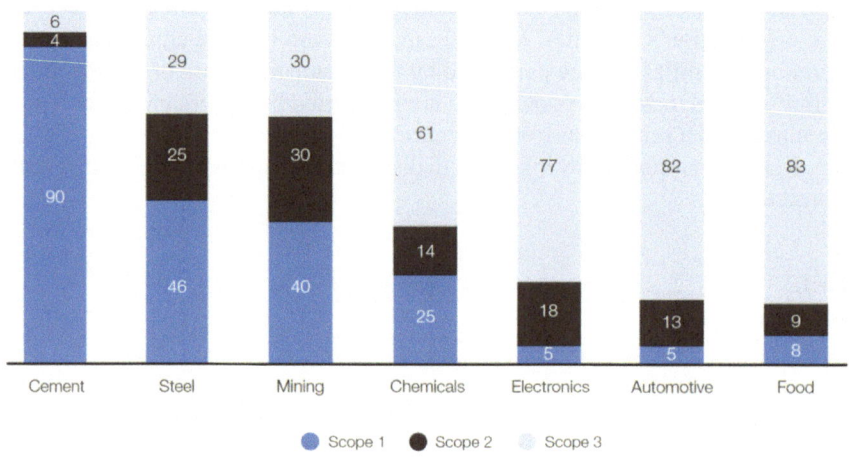

Fig. 1.3 Percentage of Scope 1, 2, and 3 GHG emissions across selected industries in 2019 [7]

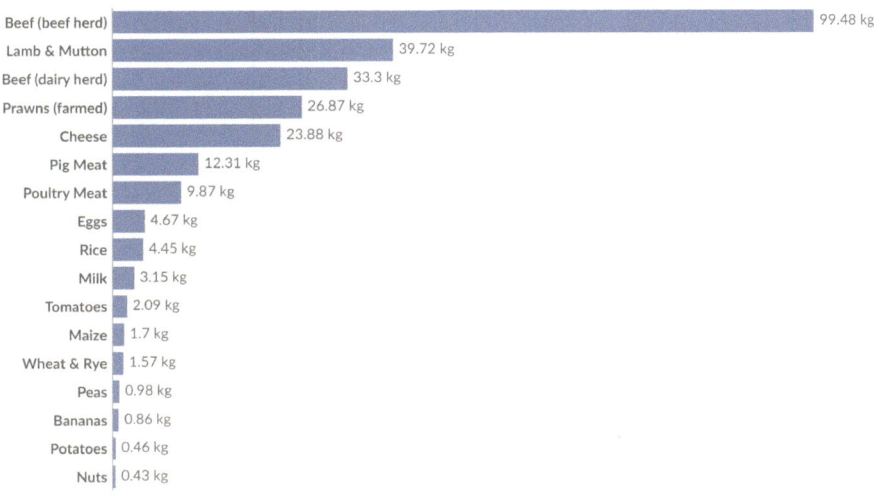

Fig. 1.4 GHG emissions per kilograms of food product [6]

Life-cycle assessment (LCA) is the concept used to estimate the GHG emissions from the entire supply chain. LCA is a technique for quantifying and evaluating the resulting environmental impacts on air, water, and soil throughout the life cycle of a product or system, including raw material extraction, processing, refining, transportation, use, and disposal. Many companies and products have recently introduced environmental assessments based on LCA.

Carbon footprint is a representative example of the application of LCA. Carbon footprint means the total amount of GHG emissions in terms of carbon dioxide equivalent emissions, which is generated by the entire process of producing, distributing, consuming, and disposing of a product by a company or individual. For example, Fig. 1.4 shows research results of GHG emissions of food products based on LCA, including land use change, livestock production, feed production, food processing, transportation, and packaging.

1.4 FuelEU Maritime and 2023 IMO GHG Strategy

Third, to qualify as a green alternative fuel, how much greenhouse gas emissions must be reduced? The ultimate answer to this question should be zero, because only then can we get to what people call a net-zero GHG world. However, we can't achieve the zero emissions overnight, so the question has to shift to what steps we need to take and by when. This is an ongoing question and we don't have a global answer yet. Instead, we can look at the current discussions to get an idea of where we might be headed.

In 2023, the European Council adopted a new regulation, FuelEU Maritime, to decarbonize the maritime sector. FuelEU Maritime will be implemented from Jan 1, 2025, and sets requirements on the GHG intensity (GHG emissions per energy, g_{CO_2eq}/MJ) by ships trading within EU or European Economic Area. The baseline is the average intensity in 2020 of 91.16 g_{CO_2eq}/MJ. The reduction target starts with 2% in 2025, and reaches to 80% by 2050, as shown in Table 1.4.

In July 2023, the International Maritime Organization (IMO) held the 80th Marine Environment Protection Committee (MEPC) and adopted the 2023 IMO Strategy on Reduction of GHG Emissions from Ships on International Voyages. The 2023 IMO GHG Strategy addresses the weaknesses of the 2018 Initial IMO GHG Strategy and includes more stringent targets and refinements. The 2023 IMO GHG Strategy sets out the strengthened decarbonization target and directions for reducing GHG emissions from international shipping in the future, and it is expected that specific methodologies will continue to be introduced, revised, and reviewed in the coming years. Table 1.5 shows the highlights of MEPC 80 and 2023 IMO GHG Strategy, compared to the existing regulations.

Figure 1.5 clearly shows that international shipping is facing much tougher GHG reduction targets. The 2023 IMO Strategy set the goal of achieving net-zero greenhouse gas emissions by 2050, and introduced indicative checkpoints to reduce the total annual GHG emissions from international shipping by at least 20% (striving 30%) by 2030 and 70% (striving 80%) by 2040, compared to 2008. In addition, the 2023 IMO GHG Strategy specified that the emissions of methane and nitrous oxide from ships should be considered for mid-term GHG reductions.

2023 IMO GHG Strategy also specified that LCA-based Well-to-Wake (WtW) GHG emissions from the production to use of marine fuels should be considered. WtW refers to the entire process for a fuel, including Well-to-Tank (WtT), which

Table 1.4 GHG intensity requirements for ships in FuelEU Maritime

Year	Reduction rate (%)	GHG intensity limit (g_{CO_2eq}/MJ)
2020	Reference value	91.16
2025–2029	2	89.34
2030–2034	6	85.69
2035–2039	14.5	77.94
2040–2044	31	62.90
2045–2050	62	34.64
2050–	80	18.23

Table 1.5 Key points of MEPC 80 and 2023 IMO GHG strategy

	Before 2023 (2018 initial IMO strategy)	2023 IMO GHG strategy and MEPC.1/Circ.905
GHG reduction target	• Reduce the total annual GHG emissions by at least 50% by 2050, compared to 2008	• Net-zero GHG emissions by or around, i.e. close to, 2050 • Reduce total annual GHG emissions by at least 20%, striving 30% by 2030 • Reduce total annual GHG emissions by at least 70%, striving 80% by 2040
Carbon intensity	• Reduce carbon intensity by at least 40% by 2030 compared to 2008 • Reduce carbon intensity by at least 70% by 2050 compared to 2008	• Reduce carbon intensity by at least 40% by 2030 compared to 2008
Regulated GHG	• Evaluate CO_2 emissions only	• Evaluate CO_2 equivalent GHG emissions including CO_2, CH_4, N_2O
Regulated scope	• Evaluate only the use (combustion) stage, i.e., Tank-to-Wake (TtW)	• Evaluate production to use, based on LCA, i.e. Well-to-Wake (WtW)
Criteria for biofuels		• Approve MEPC.1/Circ.905 on Interim guidance on the use of biofuels: WtW GHG reduction of at least 65% is required

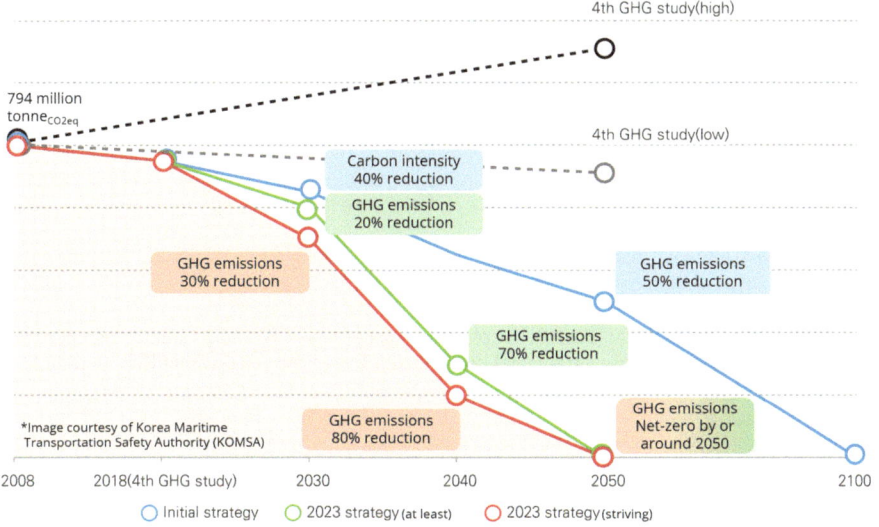

Fig. 1.5 Target of 2023 IMO GHG Strategy comparing to 2018 initial IMO Strategy

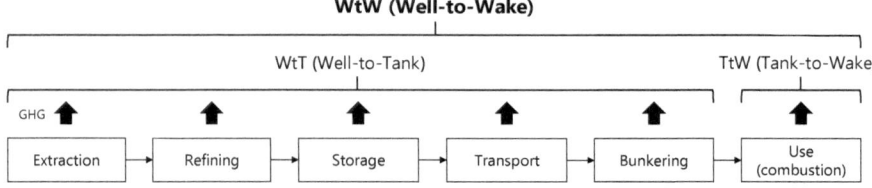

Fig. 1.6 Concept of WtT, TtW, and WtW

refers to the process of producing and transporting fuel from an oil or gas field (Well) to the ship's fuel tank (Tank), and Tank-to-Wake (TtW), which refers to the use the fuel from the fuel tank to the ship, as shown in Fig. 1.6.

The purpose of this book is to provide an introduction to alternative fuels for environmentally friendly ships for the decarbonization of shipping. Chapter 2 introduces the historical context of global warming and climate change that has led to the need for alternative fuels. Chapter 3 covers the history of regulations for maritime decarbonization and how this will change in the future through the 2023 IMO GHG Strategy. Chapters 4–6 discuss the representative alternative fuels being considered for ships, and detail how they are likely to be regulated in the future. LNG (Chap. 4), hydrogen and ammonia (Chap. 5), biofuels and e-fuels (Chap. 6) are covered. Chapter 7 describes carbon capture, utilization, and storage (CCUS) technologies, which are now becoming an important part of the alternative fuel discussion.

References

1. GHG Protocol, Standards & Guidance (2001). https://ghgprotocol.org/standards-guidance. Accessed 30 Oct 2024
2. IPCC AR4, Climate change 2007: the physical science basis. Contribution of Working Group I to the Fourth Assessment Report of the Intergovernmental Panel on Climate Change (Cambridge University Press, Cambridge, 2007)
3. IPCC AR5, Climate change 2013: the physical science basis. Working Group 1 (WG1) Contribution to the Intergovernmental Panel on Climate Change 5th Assessment Report (Cambridge University Press, Cambridge, 2013)
4. IPCC AR6, Summary for policymakers. Climate Change 2023: Synthesis Report, Contribution of Working Groups I, II and III to the Sixth Assessment Report of the Intergovernmental Panel on Climate Change (2023)
5. Karen Hills, The basics of carbon markets and trends: something to keep an eye on (2022). https://csanr.wsu.edu/the-basics-of-carbon-markets-and-trends/. Accessed 30 Oct 2024
6. Our World in Data, Greenhouse gas emissions per kilogram of food product (2018). https://ourworldindata.org/grapher/ghg-per-kg-poore. Accessed 30 Oct 2024. J. Poore, T. Nemecek, Reducing food's environmental impacts through producers and consumers. Science **360**(6392), 987–992 (2018)

7. World Economic Forum, Taking responsibility and harnessing collaboration: the keys to unlocking scope 3 emissions reduction (2019). https://www.weforum.org/agenda/2023/06/tak ing-responsibility-and-working-together-the-keys-to-unlocking-scope-3-emissions/. Accessed 30 Oct 2024

Chapter 2
Climate Change and Global Warming

Abstract Climate change and global warming are primarily driven by anthropogenic greenhouse gas (GHG) emissions from human activities. In the early days, there were debates that climate change was a conspiracy, or exaggerated. Over time, evidence supporting human-induced climate change has become overwhelming, leading to a broad scientific consensus on anthropogenic global warming. The Kyoto Protocol (1997) marked the first global effort to mitigate GHG emissions but proved limited in its effectiveness, primarily due to the absence of binding commitments from major emitters and the rapid industrial growth in developing countries. The Paris Agreement (2015) introduced a more flexible, bottom-up framework, requiring all parties to set Nationally Determined Contributions (NDCs) to reduce emissions in alignment with global temperature targets. Despite progress in international climate policy, global energy consumption remains heavily dependent on fossil fuels. Furthermore, the GHG emissions from international shipping are the eighth largest source of GHG emissions but were not included in the national GHG reduction targets. This was the starting point for the IMO to develop GHG reduction strategies for ships.

Keywords Climate change · Global warming · Greenhouse gases (GHG) · Kyoto Protocol · Paris Agreement · World energy consumption

2.1 Climate Change and Global Warming

Climate change refers to long-term shifts in the Earth's temperature and weather patterns, due to the influence of various natural and anthropogenic factors. Especially after the nineteenth century, human activities emitting anthropogenic GHGs have been focused as the main driver of climate change. In recent years, the issue of climate change has become a serious problem that can be felt by ordinary citizens, not only by scientists and environmental organizations. Heat waves, heavy rains, and various extreme weather events are becoming more frequent, and the damage they cause is increasing.

Substances in the atmosphere that can absorb energy emitted from the Earth and return it to the Earth are called GHGs. Because of this phenomenon, the Earth's temperature can be maintained and it is called as the greenhouse effect, as shown in Fig. 2.1. Basically, the greenhouse effect is a grateful effect that humans can live in the Earth. However, as human activities have greatly increased GHG emissions after the Industrial Revolution, the strengthened greenhouse effect is causing severe climate problems, which is the global warming phenomenon we are currently experiencing.

The concept of GHGs and global warming has been around for a long time. In 1896, Swedish physicist and chemist Svante Arrhenius (Fig. 2.2, this is the man who gave us the Arrhenius equation and the definition of an acid–base, and won the Nobel Prize in Chemistry in 1903) published a paper claiming that human emissions of carbon dioxide could strengthen the Earth's greenhouse effect, causing global warming. At the time, however, many people believed that the amount of carbon dioxide emitted by humans was too small to have a global impact.

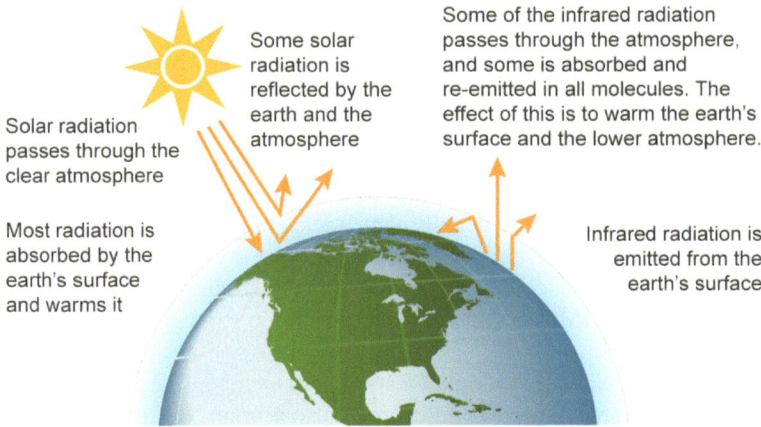

Fig. 2.1 Greenhouse effect caused by greenhouse gases

Fig. 2.2 Svante Arrhenius (1859–1927)

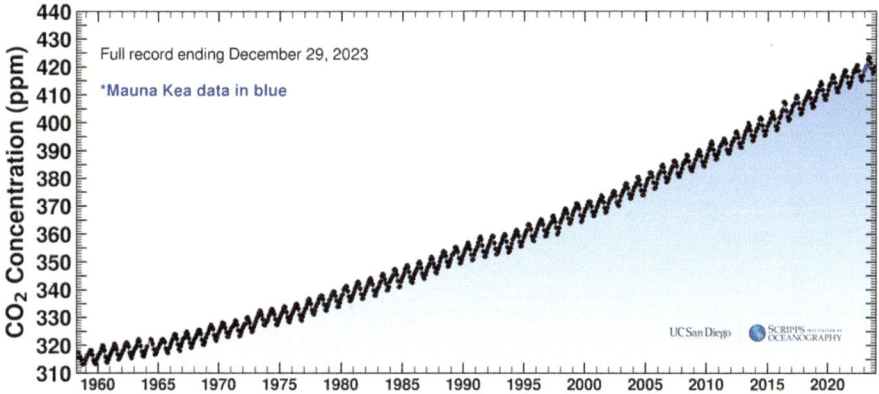

Fig. 2.3 Keeling curve showing the increasing carbon dioxide concentration [3, 5]

In 1958, American scientist Charles D. Keeling (1928–2005) began a study measuring atmospheric carbon dioxide concentrations at the Mauna Loa Observatory in Hawaii. The results of his research led to the publication of the Keeling Curve in 1961, which showed that the concentration of carbon dioxide had been continuously increasing as shown in Fig. 2.3, reigniting the climate change debate.

2.2 The Kyoto Protocol (1997)

In 1979, the World Climate Conference was organized to discuss about the climate change. In 1988, the Intergovernmental Panel on Climate Change (IPCC) was formed, and the global discussion on climate change began in earnest. In 1992, the United Nations adopted the United Nations Framework Convention on Climate Change (UNFCCC), and in 1997, the Kyoto Protocol, the world's first global agreement to reduce GHG emissions, was adopted. The Kyoto Protocol identifies the developed countries as Annex I Parties, which are required to reduce GHG emissions, and assigns them a target of reducing greenhouse gas emissions by 5% compared to 1990 levels.

However, the Kyoto Protocol was not successful in reducing global GHG emissions. According to the UNFCCC, in 2009, GHG emissions from Annex I countries decreased by more than 10% compared to 1990, but global GHG emissions actually increased by about 38%. The first reason for this was the rapid growth of China, India, and other non-Annex I countries as shown in Fig. 2.4, which were classified as developing countries in 1997 and were not obligated to reduce their GHG emissions. The second reason was that the United States, the world's largest emitter of GHGs in 1990s, did not ratify the Kyoto Protocol. The Kyoto Protocol then ended without much success, with Canada withdrawing from the agreement in 2011 and Japan and

Russia refusing to commit to reductions during the second extension period that
began in 2013. Figure 2.5 shows the parties in the Kyoto Protocol.

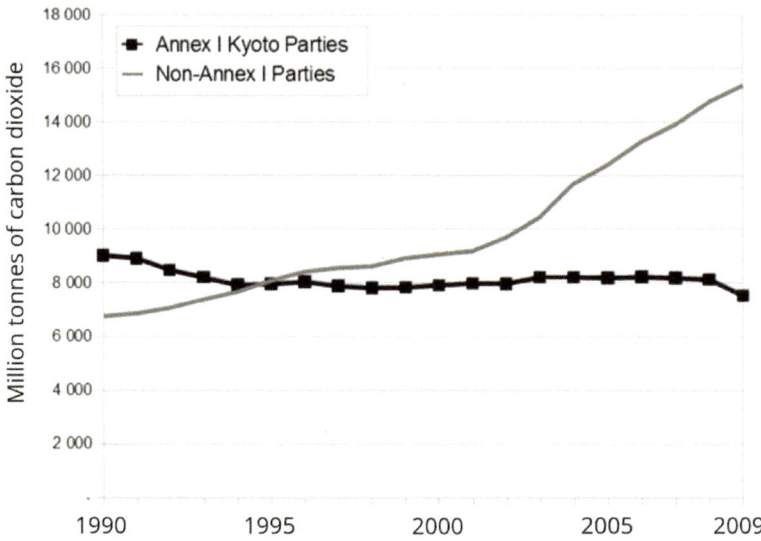

Fig. 2.4 Annual carbon dioxide emissions from fuel combustion between 1990 and 2009 for the
Kyoto Annex I and non-Annex I Parties [2]

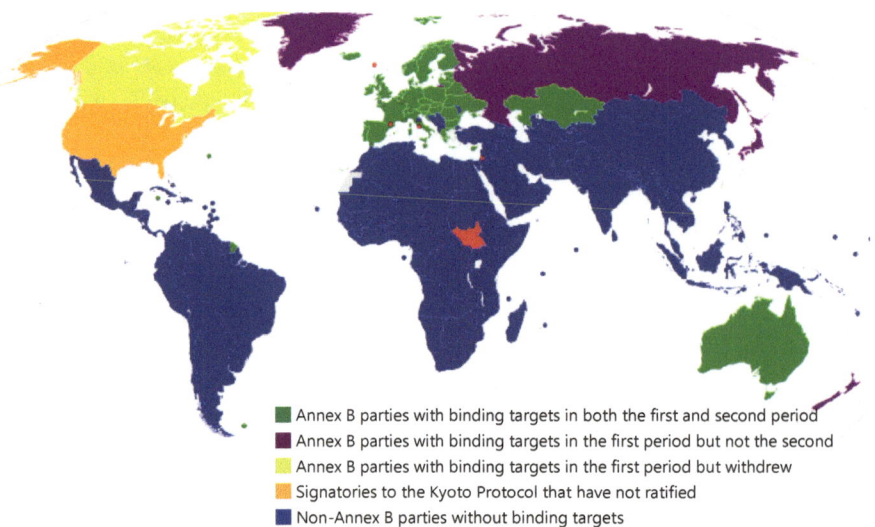

Fig. 2.5 Parties in the Kyoto Protocol

2.3 Scientific Consensus on Anthropogenic Global Warming

You may have heard claims that climate change is a conspiracy, or exaggerated. In the early days of the climate change debate, there were many counterarguments that denied global warming, or that global warming was a part of the Earth's natural climate change cycle, not caused by human activities. Climate change skepticism, the idea that the issue of climate change and global warming has been exaggerated by certain parties has been around since before the 1990s.

To resolve this controversy, paleoclimatology, which uses indirect sources of past climate information to estimate the climate of ancient times, has grown. One such method is the use of ice cores from Antarctica. Snow is repeatedly accumulated and compressed without melting, and becomes ice. When we get a deep ice core, air bubbles containing the past atmosphere exist along with the ice, so it is possible to estimate the composition of the past atmosphere by analyzing them.

A number of researches have been conducted to estimate past atmospheric conditions on the Earth, and the results of these analyses have been published and cross-checked by different research teams. Based on these analyses, it is estimated that the current increase in carbon dioxide concentration is extremely rapid compared to the past, even if we look at the estimated record going back 800,000 years as shown in Fig. 2.6.

As a result of these studies, the IPCC's Sixth Assessment Report in 2023 stated the following [6]:

It is unequivocal that human influence has warmed the atmosphere, ocean and land. Widespread and rapid changes in the atmosphere, ocean, cryosphere, and biosphere have occurred.

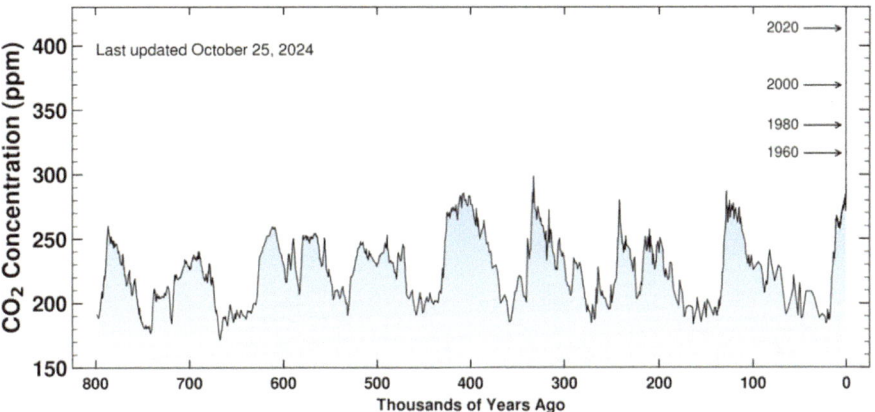

Fig. 2.6 CO_2 concentrations over the last 800,000 years [3–5]

Table 2.1 NDC targets of several countries (UNFCCC)

Australia	43% reduction by 2030 based on 2005, net-zero by 2050
Brazil	37%, 50% reduction by 2025, 2030 based on 2005, net-zero by 2050
Canada	40–45% reduction by 2030 based on 2005
China	65% reduction by 2030 based on 2005, net-zero before 2060
India	45% reduction by 2030 based on 2005
Japan	46% reduction by 2030 based on 2013, net-zero by 2050
Russia	30% reduction by 2030 based on 1990
UK	At least 68% reduction by 2030 base on 1990
USA	50–52% reduction by 2030 based on 2005
EU	At least 55% reduction by 2030 based on 1990

In other words, we have now reached a scientific consensus where the majority of scientists recognize anthropogenic climate change as the majority theory.

2.4 Paris Agreement (2015)

The Paris Agreement is a legally binding international treaty on climate change. The Paris Agreement has a long-term temperature goal to hold "the increase in the global average temperature to well below 2 °C above pre-industrial levels" and to pursue efforts "to limit the temperature increase to 1.5 °C above pre-industrial levels." It was adopted by 196 Parties (195 Parties and the European Union) at the UNFCCC's 21st Conference of the Parties (COP 21) in Paris in 2015, and as an agreement without an end date, it entered into force on November 4, 2016. The United States withdrew from the agreement in 2020, but rejoined in 2021.

While the Kyoto Protocol had only 40+ parties with obligations to reduce GHG emissions, the Paris Agreement requires all Parties to set Nationally Determined Contributions (NDCs), which set their own GHG reduction targets based on their own circumstances and capabilities as shown in Table 2.1. The NDCs should be set every five years and the information about the implementation should be provided. Many countries have now declared their target and developing long-term strategies and legislation to achieve the goal.

2.5 World Energy Consumption and GHG Emissions

While the Paris Agreement represents a global agreement to reduce greenhouse gas emissions, it is not an easy goal to achieve. Figure 2.7 shows global primary energy consumption from 2000 to 2021, and the share of use by energy source. Primary energy means the energy obtained directly from nature without further conversion, including coal, oil, natural gas, hydroelectric, nuclear, solar, and wind. There have been only two periods in the last 20 years when global energy consumption has declined. Only in 2008 due to the global financial crisis, and in 2020 due to the COVID-19 pandemic, we can see a temporary drop in energy use. Otherwise, energy consumption has steadily increased each year.

When checking the energy use, it is clear that we still get the majority of energy from fossil fuels. Although the share of fossil fuels has decreased from the past, coal (26.9%), oil (31.0%), and natural gas (24.4%) accounted for 82.3% of total energy use in 2021, meaning that we still get more than 80% of energy from fossil fuels. The share of renewable energy is steadily increasing, but even including hydropower, it only accounts for 13% of total energy use in 2021.

This is the reason why people are currently interested in alternative fuels. With the development of various applied technologies, it seems that the world's energy consumption would not decrease in the future. Under these circumstances, the need for alternative fuels that can replace fossil fuels will increase.

Figure 2.8 shows the total GHG emissions by major emitters in 2020 including the Land Use, Land Use Change, and Forestry (LULUCF) sector of the GHG inventory [7]. Forests, grasslands, and wetlands can absorb and store carbon dioxide, and can therefore be credited with GHG mitigation. On the other hand, deforestation could result in increased GHG emissions in the LULUCF sector.

Notably, international shipping is the eighth largest source of GHG emissions in the world. However, ships on international voyages emit GHGs on the high seas (public waters), are not in a country's territorial waters. This creates a blind spot for

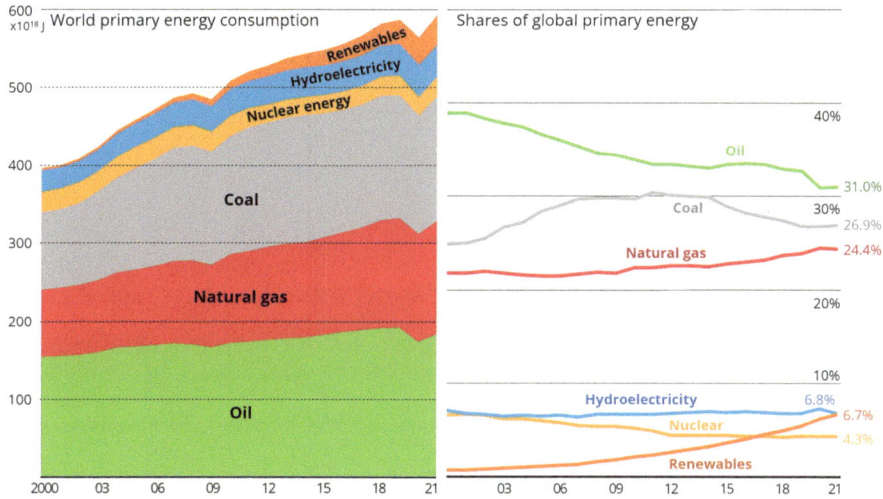

Fig. 2.7 World primary energy consumption and shares by energy source [1]

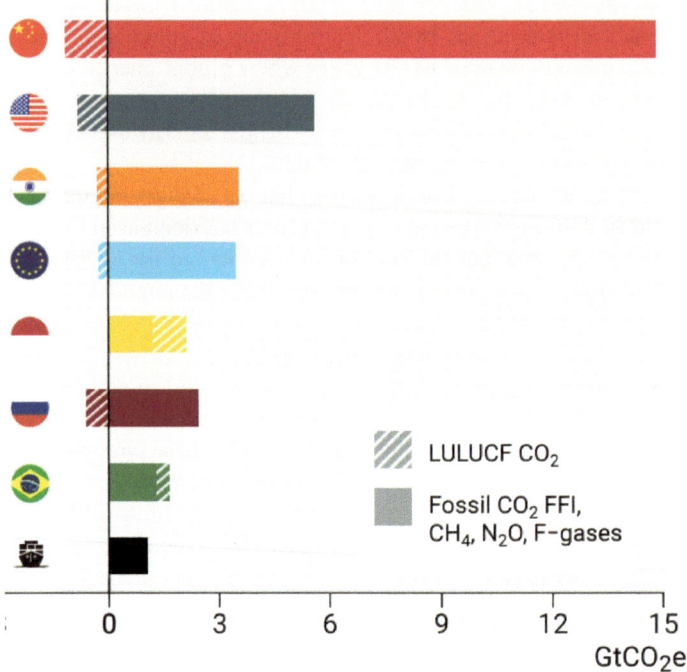

Fig. 2.8 Total GHG emissions of major emitters in 2020 [6]

national GHG reduction targets. This is the main reason why the IMO is developing and discussing GHG reduction strategies for ships sailing internationally. In Chap. 3, we will therefore discuss GHG reduction regulations for international shipping in more detail.

References

1. BP, Statistical Review of World Energy (2022)
2. IEA, *CO₂ Emissions from Fuel Combustion 2011* (OECD Publishing, 2011). https://doi.org/10.1787/co2_fuel-2011-en
3. C.D. Keeling et al., Exchanges of atmospheric CO_2 and 13CO₂ with the terrestrial biosphere and oceans from 1978 to 2000. I. Global Aspects, SIO Reference Series, No. 01-06, Scripps Institution of Oceanography, San Diego (2001), p. 88
4. D.M. Lüthi et al., High-resolution carbon dioxide concentration record 650,000–800,000 years before present. Nature **453**, 379–382 (2008)
5. Scripps Institution of Oceanography, The keeling curve (2024). https://keelingcurve.ucsd.edu/. Accessed 30 Oct 2024
6. IPCC AR6, Summary for policymakers. Climate change 2023: synthesis report, contribution of working groups I, II and III to the sixth assessment report of the intergovernmental panel on climate change (2023)
7. UNEP, Emissions Gap Report 2022: The Closing Window—Climate crisis calls for rapid transformation of societies (2022). https://www.unep.org/emissions-gap-report-2022

Chapter 3
IMO GHG Strategy

Abstract As a specialized agency of the United Nations (UN), the International Maritime Organization (IMO) plays a key role in regulating maritime activities on the high seas which are beyond the sovereignty and jurisdiction of any particular state. The IMO's Marine Environment Protection Committee (MEPC) adopted the international convention for the prevention of MARine POLlution from ships (MARPOL) in 1978, and revised it in 2011 to include regulations to reduce CO_2 emissions reduction from international shipping in response to global climate concerns. To assess the carbon intensity of ships, the IMO introduced the energy efficiency design index (EEDI) for newbuild ships, the energy efficiency existing index (EEXI) for existing ships, and the carbon intensity indicator (CII), which assesses the annual operational performance of international shipping. The IMO has strengthened greenhouse gas reduction targets, aiming for net-zero emissions by 2050. Future regulations are expected to consider the well-to-wake (WtW) GHG emissions intensity based on lifecycle assessments, including greenhouse gases other than CO_2.

Keywords IMO · MEPC · MARPOL · EEDI · EEXI · CII · Carbon intensity

3.1 Territorial and High Seas

Figure 3.1 shows the concept of territorial sea and high seas. Territorial sea is the sea over which a coastal state has sovereignty and jurisdiction, now up to 12 nautical miles (1 nm = 1.852 km) from the baseline. In other words, territorial sea means how far above the sea can be recognized as a country's land. Of course, this distance has long been disputed. From the fourteenth to seventeenth centuries, there were many different ideas about the limit of the territorial seas of 100, 60, and 50 nm, range of cannon shot or range of eyesight. In the eighteenth century, three nautical miles (1 nm = 1.852 km) from a land reference line was suggested, and supported by some countries.

In 1973 at the Third United Nations (UN) Conference on the Law of the Sea, the UN Convention on the Law of the Sea was adopted, and the limit of the territorial sea

© The Author(s), under exclusive license to Springer Nature Switzerland AG 2025   21
Y. Lim, *Alternative Fuels for Environmentally-Friendly Ships*,
SpringerBriefs in Applied Sciences and Technology,
https://doi.org/10.1007/978-3-031-85082-0_3

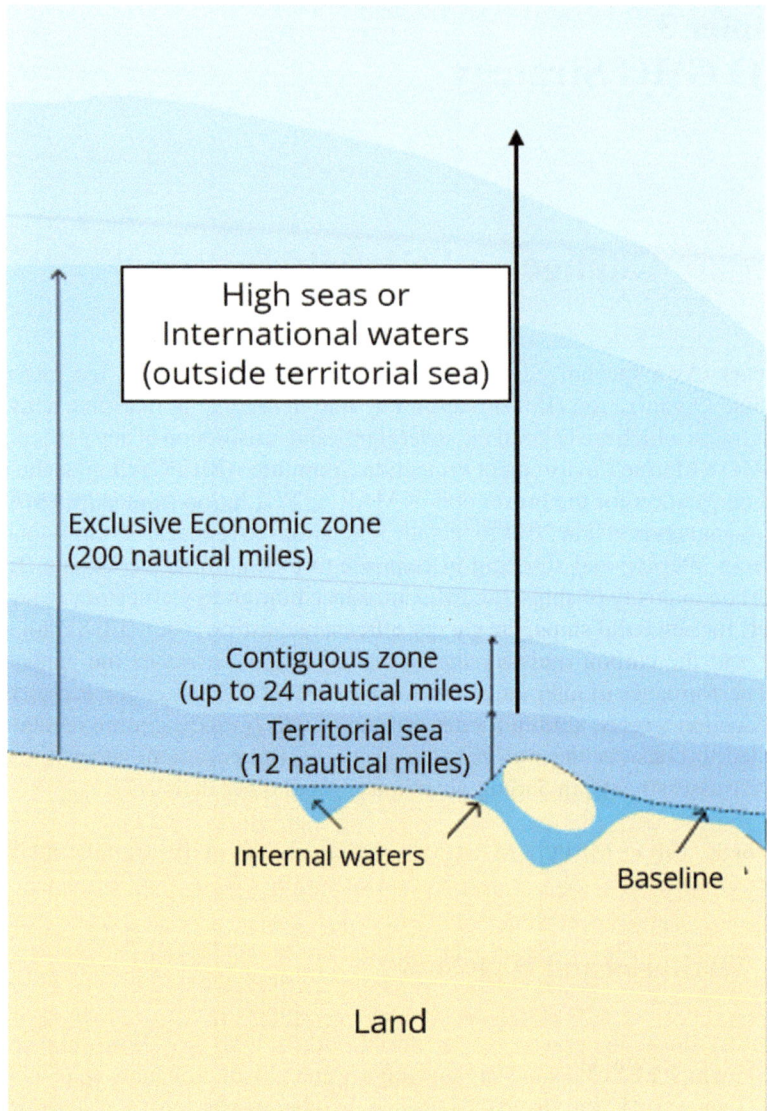

Fig. 3.1 Territorial sea and high seas/international waters [1]

was set at most 12 nm from the baseline. If the territorial sea of one state overlaps with another state's territorial sea within 12 nm, the border can be adjusted by mutual agreement.

On the other hand, the seas that are not included in the territorial sea of a state, are called the high seas (or international waters). The high seas are subject to the doctrine of *mare liberum* (freedom of the seas), and are open to all ships. Ships sailing on the high seas are generally under the flag state, which is the jurisdiction under whose laws the ship is registered, like a person's nationality. The flag state has the authority to enforce regulations on ships registered under its flag, including the prevention of pollution.

If a ship sailing on the high seas is subject to the regulations of its flag state, conflicts of interest may arise if the regulations vary from state to state. To solve these conflicts, international discussions are required to common regulations for ships on international voyages. The International Maritime Organization (IMO) was created for this purpose.

3.2 IMO MARPOL

The UN has specialized agencies for major international issues in specific areas that require expertise, such as economic, social, cultural, education, and health. Examples of UN specialized agencies are the International Monetary Fund (IMF), World Health Organization (WHO), and so on. The International Maritime Organization (IMO) is also a specialized agency of the UN and is responsible for everything that happens on the world's oceans including maritime safety, security, and environmental issues. The IMO discusses and enforces regulations that apply to ships on the high seas.

The IMO has a number of committees and specialized groups under its umbrella as shown in Fig. 3.2. One of them is the Marine Environment Protection Committee (MEPC), which conducts various discussions to prevent environmental pollution at sea. The MEPC is primarily responsible for deliberating on matters relating to the prevention and control of marine pollution by ships and for the adoption and revision of international conventions for ships on international votages.

One of the major conventions adopted by the IMO MEPC is the international convention for the prevention of MARine POLlution from ships (MARPOL), which is referred to frequently in this book. MARPOL was first discussed in 1973 to prevent marine pollution from oil spills from tankers, and was adopted in 1978. Since then, it has been revised as necessary to regulate various pollutants other than oil, adding

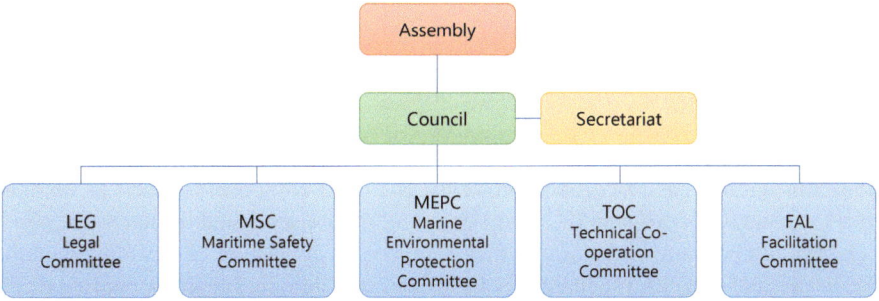

Fig. 3.2 Major committees in IMO

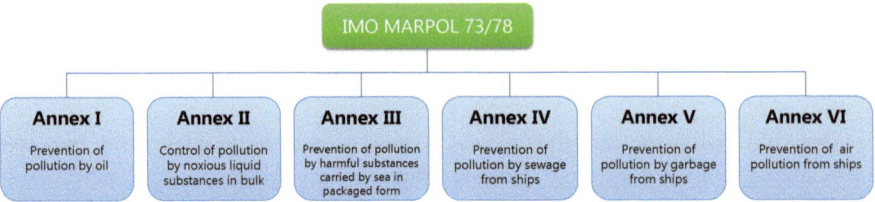

Fig. 3.3 Annexes in MARPOL

annexes and defining specific regulations. As of 2024, there are six annexes, as shown in Fig. 3.3.

- ANNEX I: Prevention of pollution by oil and oily waters. Enforced in 1983.
- ANNEX II: Control of pollution by noxious liquid substance in bulk. Enforced in 1983.
- ANNEX III: Prevention of pollution by harmful substances carried by sea in packaged form. Enforced in 1992.
- ANNEX IV: Prevention of pollution by sewage from ships. Enforced in 2003.
- ANNEX V: Prevention of pollution by garbage from ships. Enforced in 1988.
- ANNEX VI: Prevention of air pollution from ships. Enforced in 2005.

In particular, MARPOL Annex VI is closely related to the recently mentioned issues of environmentally friendly ships. MARPOL ANNEX VI originally consisted of regulations for air pollutants such as nitrogen oxides (NO_x), sulfur oxides (SO_x), volatile organic compounds, and ozone depleting substances. Later, as the issue of climate change due to greenhouse gas emissions and global warming became a global debate, the 62nd MEPC in 2011 revised MARPOL Annex VI, to include regulations to reduce CO_2 emissions from international voyage ships.

3.3   Understanding HFO and Fuels for a Ship

Most of the energy we use today comes from fossil fuels, and ships, which require large amounts of energy for propulsion, are no exception. The largest portion of use of fossil fuels is the oil (petroleum), and they are categorized into various petroleum products. Refineries produce petroleum products by distillation of crude oil as shown in Fig. 3.4. Distillation is the process of separating the components of a mixture using the differences in boiling points of the components. At atmospheric pressure, the light natural gas (NG) and petroleum gas (PG) components, such as methane (C1), ethane (C2), propane (C3), and butane (C4), are separated from a crude oil. At a higher temperature by heating, other components consisting of gasoline, naphtha, kerosene, and diesel are sequentially vaporized, and liquefied again by cooling it. The petroleum products obtained by boiling and then cooling again are called distillate oils. Among the distillate oils, the resulting diesel components are then reblended to

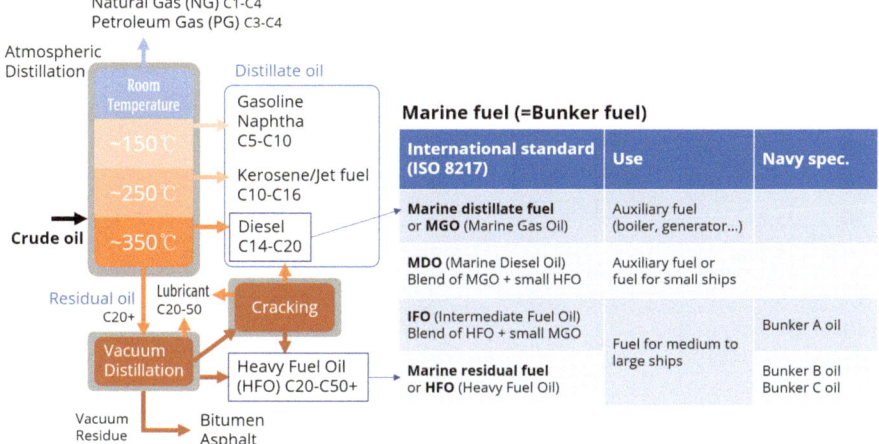

Fig. 3.4 Marine fuels for ships

meet the criteria of international marine fuel oil standards, which is called as marine gas oil (MGO).

The oil remaining after atmospheric distillation is called residual oil, and additional petroleum products can be produced from the residual oil by vacuum distillation. After further separation of residual bitumen (used in the production of road asphalt), heavy fuel oil (HFO) is produced, also known as bunker C oil. HFO and intermediate fuel oil (IFO), which is a blend of HFO and MGO, have traditionally been used as fuels for medium and large ships.

HFO is a mixture of heavy hydrocarbons, having carbon numbers from C20 to greater than C50. HFO has high heating value, so it has been widely used for large ships to use large amount of energy at a low cost. However, due to the high impurities in the residual oil, HFO emits a lot of air pollutants such as SO_x, NO_x particulate matter (PM), and greenhouse gases such as carbon dioxide also.

3.4   History of the Regulation of GHG Emissions from International Shipping

CO_2 emissions from international shipping accounted for approximately 3% of global CO_2 emissions in 2018, and emissions have been steadily increasing as shown in Table 3.1. With the adoption of the UNFCCC in 1992 and the Kyoto Protocol in 1997, the need to regulate carbon emissions from ships was recognized. In 2000, the IMO published the "IMO GHG Study," its first report on GHG emissions from ships on international voyages, and specific discussions begin on how to reduce CO_2 emissions from ships.

Table 3.1 CO_2 emissions from total shipping 2012–2018 [2]

	Global anthropogenic CO_2 emissions (million ton)	CO_2 emissions from total shipping (million ton)	Percentage (%)
2012	34,793	962	2.76
2013	34,959	957	2.74
2014	35,225	964	2.74
2015	35,239	991	2.81
2016	35,380	1,026	2.90
2017	35,810	1,064	2.97
2018	36,573	1,056	2.89

In 2009, the second IMO GHG study was published, and in 2011 at the 62nd MEPC, MARPOL Annex 6 was revised to include regulations for the reduction of CO_2 emissions from ships on international voyages for the first time. The concept of carbon intensity, which is the amount of carbon dioxide emitted per unit transportation work, was defined to regulate CO_2 emissions. The energy efficiency design index (EEDI) was introduced to assess the carbon intensity of a newbuild ship (see Sect. 3.5) [3]. The reduction target was set at 30+% by 2025, with mandatory application starting from 2013.

The third IMO GHG study was published in 2014, and the Paris agreements was adopted in 2015. In 2018, the IMO adopts the "initial IMO GHG strategy," which sets an even more ambitious target of a 40% reduction in carbon dioxide emissions intensity from international shipping by 2030 compared to 2008 levels, and striving for a 70% reduction by 2050 [4]. The fourth IMO GHG study was published in 2020, and at the 76th MEPC meeting in 2021, a revision was adopted to apply the energy efficiency existing index (EEXI) to existing ships as well as newbuild ships, and to introduce the carbon intensity indicator (CII) as a measure for ships in actual operation [5]. The EEXI and CII began to be applied from 2023.

In July 2023, the IMO 80th MEPC meeting adopted the "2023 IMO GHG Strategy," which significantly strengthened the goal of achieving net-zero GHG emissions by or around, i.e. close to 2050 [12]. Indicative checkpoints were set to reduce the total annual GHG emissions from international shipping by at least 20% (striving 30%) and 70% (striving 80%) by 2030 and 2040, respectively. In addition, the target substances for reduction would be expanded from just CO_2 to include other GHGs. The LCA would be introduced to assess the well-to-wake (WtW) GHG emissions intensity, including GHG emissions from the production phase (WtT) as well as the use phase (TtW). The specific methodologies will be discussed by the MEPC in 2024 and beyond. That is, now the era of truly decarbonized ship begins, where all GHG emissions, not just carbon dioxide, are evaluated from the production to the final use of a fuel.

3.5 Carbon Intensity for a Ship

Carbon intensity (or emission intensity) is a concept used to assess GHG emissions at different scales, meaning the amount of GHG emissions per unit of amount or specific activity. For example, the energy industry defines carbon intensity as "the amount of greenhouse gases generated per unit of energy produced" because its primary product is energy. In this case the units of g_{CO_2eq}/kWh or g_{CO_2eq}/MJ are used to indicate how many carbon dioxide equivalent GHGs are emitted to produce 1 kWh or 1 MJ of energy. CO_2-equivalent emissions refer to the conversion of all greenhouse gas emissions into carbon dioxide emissions, reflecting the global warming potential (GWP) of the greenhouse gases (see Sect. 1.2).

In the case of a ship, the primary purpose is to transport cargo, so the transport work is used instead of energy. That is, the definition of carbon intensity for a ship can be written as follows.

Carbon intensity of a ship is defined as GHG emissions per transport work.

There are many different ways to define a ship's transport work, but the IMO uses the ton-mile measurement, which represents the transport of one ton of cargo over a distance of one mile. This is a common metric used for transportation by rail and aircraft. The concept itself is the same for a ship, but, unlike land, the unit of distance for a ship is the nautical mile (1 nm = 1.852 km), rather than the land mile (1 mile = 1.6 km). That is, the carbon intensity of a ship can be rephrased as follows.

Carbon intensity of a ship is defined as GHG emissions for transporting one ton of cargo over one nautical mile.

3.6 Index and Indicator for the Carbon Intensity of a Ship

The energy efficiency design index (EEDI), first introduced in 2011 to estimate a ship's carbon intensity [3]. The exact mathematical definition is much more complex, but in an easy-to-understand form it is defined as Eq. 3.1.

$$\text{EEDI} = \frac{CO_2 \text{ emissions}(g_{CO_2})}{\text{Cargo weight(t)} \cdot \text{Trasport distance(nm)}} \tag{3.1}$$

The EEDI estimates only carbon dioxide emissions from a ship. A ship with an EEDI of 1 g_{CO_2}/(t.nm) emits 1 g of carbon dioxide per nautical mile of cargo transported, so the higher the EEDI the more carbon dioxide the ship emits. The EEDI only applies to internationally operated ships of 400 gross tonnage and above, sailing on the high seas. The EEDI also only applies to a new ship built in 2013 or later, or to an existing ship that has undergone a major conversion. Because the EEDI applies to a newbuild ship, it is not based on actual data but is derived from design specifications, meaning that the targeted reductions need to be already achieved in the pre-design phase.

Because the EEDI is only applied to newbuild ships, existing ships were able to operate free of CO_2 emissions regulations even after the application of the EEDI started in 2013. To solve this problem, the energy efficiency existing index (EEXI) was introduced at the 76th MEPC in 2021, and existing ships also have been subject to CO_2 emissions regulation from 2023. The EEXI is the same concept as the EEDI, but differs in that it applies to existing ships rather than new ships [5−7].

The carbon intensity indicator (CII), also introduced in 2021 with EEXI, is essentially the same meaning of carbon intensity, the CO_2 emissions per transport work. However, CII is an indicator calculated from actual the operational data of a ship based on the actual fuel consumption, cargo weight, and transport distance, while EEDI and EEXI are indices based on the design specification on the document. CII is defined as Eq. 3.2 [8−11].

$$
\begin{aligned}
CII &= \frac{\text{Fuel consumption}(g_{Fuel}) \cdot \text{Conversion factor}(g_{CO_2}/g_{Fuel})}{\text{Cargo weight(t)} \cdot \text{Trasport distance(nm)}} \\
&= \frac{CO_2 \text{ emissions}(g_{CO_2})}{\text{Cargo weight(t)} \cdot \text{Trasport distance(nm)}}.
\end{aligned} \tag{3.2}
$$

Based on the CII, a ship has a rating from A to E depending on its level of performance. The rating indicates a major superior (A), minor superior (B), moderate (C), minor inferior (D), or inferior (E) performance level. If a ship is rated D for three consecutive years or E for one year, it has to submit a corrective action plan to show how it will achieve the required rating of C or above.

Currently several types of indicators are used for the CII, and this book introduces energy efficiency operational indicator (EEOI) and annual efficiency ratio (AER). EEOI is a carbon intensity calculated from the actual fuel consumption, cargo weight, and transport distance as defined in Eq. 3.2. In 2016 the IMO adopted the IMO Data Collection System (DCS) to collect and report the operational data of individual ships, starting from January 1, 2019, so now the data is being collected from ships. However, before 2019, some ships had data and others did not. In particular, there is often no actual cargo weight data for a ship, then the EEOI cannot be calculated. In this case, the AER can be used. The AER is not calculated from the actual cargo weight, but is based on the deadweight (DWT) of the ship, which is the maximum weight of cargo that the ship can carry as shown in Eq. 3.3.

$$
CII(\text{based AER}) = \frac{\text{Fuel consumption}(g_{Fuel}) \cdot \text{Conversion factor}(g_{CO_2}/g_{Fuel})}{\text{Deadweight(t)} \cdot \text{Trasport distance(nm)}}
$$

$$= \frac{CO_2 \text{ emissions}(g_{CO_2})}{\text{Deadweight(t)} \cdot \text{Trasport distance(nm)}}. \quad (3.3)$$

It should be noted that CII based on AER is always lower than that based on EEOI because AER uses deadweight, which is always higher than the actual cargo weight of a ship. The 2020 IMO GHG Fourth study shows that the AER is about half of the EEOI as shown in Table 3.2, because not all ships operate at full deadweight all the time.

In 2020, IMO published the 4th IMO GHG study, which shows the GHG emissions of international shipping from 1990 to 2018, based on the GHG emissions in 2008, as shown in Fig. 3.5. It shows that before 2008, the growth of GHG emissions was tightly coupled to the growth of seaborne trade, but after 2008, despite the growth of seaborne trade, carbon intensity such as EEOI and AER were decoupled from the growth of transport demand. This means that the IMO regulations for GHG reduction are having some effect. However, after 2014, as maritime trade demand grows, the total GHG emissions return to growth trend, which may explain why the MEPC had set more stringent GHG reduction targets.

As we reviewed, the EEDI, EEXI, and CII were originally defined to assess only carbon dioxide emissions from the fuel combustion phase of ships without considering other GHGs. With the adoption of the 2023 IMO GHG Strategy [12] as mentioned in Sect. 1.4, it has been discussed that the assessment of GHG emissions from ships in the future would consider WtW GHG emissions from the production process as well as the use of fuel, and GHGs other than carbon dioxide. Specific decisions on whether to revise existing indices or introduce new ones have not yet been taken, but it is expected that the IMO committees will work out the details of the implementation after 2024.

As GHG emissions regulations for international shipping are tightened toward the goal of net-zero, there is a lot of discussion about fundamentally changing the

Table 3.2 Estimates on carbon intensity and percentage changes 2018 compared to 2008 [2]

Year	EEOI $(g_{CO_2}/(\text{t.nm}))$				AER $(g_{CO_2}/(\text{DWT.nm}))$			
	Vessel-based		Voyage-based		Vessel-based		Voyage-based	
	EEOI	Change (%)	EEOI	Change (%)	AER	Change (%)	AER	Change (%)
2008	17.10	–	15.16	–	8.08	–	7.40	
2012	13.16	− 23.1	12.19	− 19.6	7.06	− 12.7	6.61	− 10.7
2013	12.87	− 24.7	11.83	− 22.0	6.89	− 14.8	6.40	− 13.5
2014	12.34	− 27.9	11.29	− 25.6	6.71	− 16.9	6.20	− 16.1
2015	12.33	− 27.9	11.30	− 25.5	6.64	− 17.7	6.15	− 16.9
2016	11.87	− 28.6	11.21	− 26.1	6.58	− 19.5	6.09	− 17.7
2017	11.67	− 30.6	10.88	− 28.2	6.43	− 20.4	5.96	− 19.5
2018	11.67	− 31.8	10.70	− 29.4	6.31	− 22.0	5.84	− 21.0

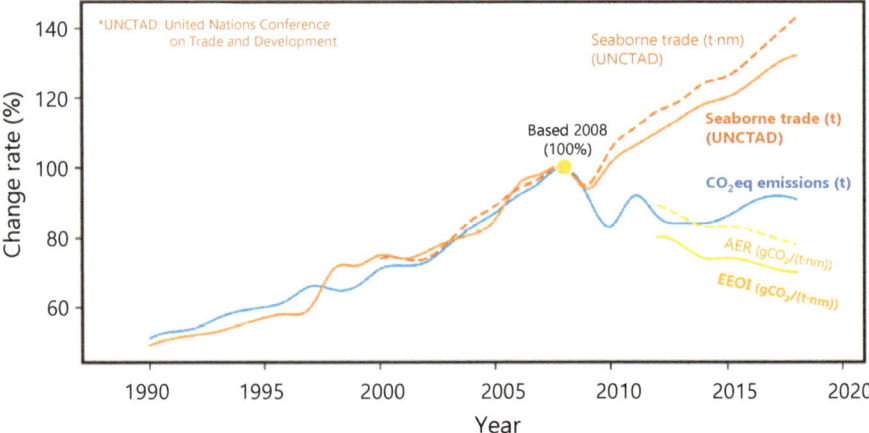

Fig. 3.5 International shipping GHG emissions indexed in 2008, voyage-based [2]

fuel in addition to efficiency improvements and energy reduction research. For this reason, the discussion of alternative fuels for ships has recently become a hot topic. From Chap. 4, we will discuss in more detail the advantages and limitations of the various fuels that have been proposed as alternative fuels.

References

1. Historicair, Maritimes zones in UNCLOS (2006). https://commons.wikimedia.org/wiki/File: Zonmar-en.svg. Accessed 30 Oct 2024
2. IMO, Fourth IMO GHG Study. International Maritime Organization (2020)
3. Resolution MEPC.203(62), Amendments to the annex of the protocol of 1997 to amend the international convention for the prevention of pollution from ships (2011)
4. Resolution MEPC.304(72), Initial IMO strategy on reduction of GHG emissions from ships (2018)
5. Resolution MEPC.333(76), 2021 guidelines on the method of calculation of the attained energy efficiency existing ship index (EEXI) (2021)
6. Resolution MEPC.334(76), 2021 guidelines on survey and certification of the energy efficiency existing ship index (EEXI) (2021)
7. Resolution mepc.335(76), 2021 guidelines on the shaft/engine power limitation system to comply with the EEXI requirements and use of a power reserve (2021)
8. Resolution MEPC.336(76), 2021 guidelines on operational carbon intensity indicators and the calculation methods (CII guidelines, G1) (2021)
9. Resolution MEPC.337(76), 2021 guidelines on the reference lines for use with operational carbon intensity indicators (CII reference lines guidelines, G2) (2021)
10. Resolution MEPC.338(76), 2021 guidelines on the operational carbon intensity reduction factors relative to reference lines (CII reduction factor guidelines, G3) (2021)

11. Resolution MEPC.339(76), 2021 guidelines on the operational carbon intensity rating of ships (CII rating guidelines, G4) (2021)
12. Resolution MEPC.377(80), 2023 IMO strategy on reduction of GHG emissions from ships (2023)

Chapter 4
Liquefied Natural Gas (LNG)

Abstract This chapter explores the role of liquefied natural gas (LNG) in global energy systems, its environmental impact, transportation issues, and potential as a marine fuel. LNG is a fossil fuel consisting primarily of methane, with smaller amounts of ethane, propane, butane, nitrogen, and carbon dioxide. A large LNG carrier is typically used because LNG's high density compared to NG makes it economical for long-distance transport, but it has the operational issues of boil-off gas (BOG) generation due to the very low storage temperature of LNG. With stricter GHG emission regulations, advanced dual-fuel gas injection engines with reliquefaction are now being used. While LNG emits less CO_2 than conventional heavy fuel oil (HFO), its GHG emissions intensity may close to that of HFO when considering engine type and methane slip. Demand for LNG as a fuel will gradually decline in the distant future because it is essentially a fossil fuel, but the use of LNG as a feedstock for alternative fuels such as hydrogen is expected to continue for a considerable period of time. In addition, alternative sources of LNG, such as biogas, can offer a more sustainable future by reducing net GHG emissions.

Keywords LNG · BOG · GHG intensity · Methane slip · Biomethane · Bio-LNG

4.1 What is LNG?

There has been a lot of debate about whether LNG is an environmentally friendly alternative fuel or not. The LNG we use today is the liquefied natural gas (LNG), which is essentially a fossil fuel. Natural gas (NG) is a mixture of gases consisting primarily of methane, typically 70–95%, with smaller amounts of other gases such as ethene, propane, butane, nitrogen, and carbon dioxide as shown in Table 4.1. The organic chemicals composed up of the carbon (C) and hydrogen (H) are called as hydrocarbons, which are the major components of fossil fuels. Methane is representative hydrocarbon combining one carbon atom with four hydrogen atoms (CH_4), and naturally emits carbon dioxide when combusted.

Table 4.1 Major composition of typical natural gas

Name	Composition (mole fraction) (%)	Molecular formula	Boiling point (°C, at 1 atm)
Methane	70–95	CH_4	− 161.5
Ethane	~ 4–6	C_2H_6	− 88.6
Propane	~ 2–3	C_3H_8	− 42.1
Butane	~ 1–3	C_4H_{10}	(n-butane) − 0.5 (i-butane) − 11.7
Nitrogen	~ 1	N_2	− 195.8
Carbon dioxide	~ 1	CO_2	− 78.6

While LNG has clear limitations as a fossil fuel, it also has advantages that cannot be ignored at this time. In addition, the alternative fuel supply chains currently being discussed are often modeled after the structure of LNG supply chains, so an understanding of LNG is essential to understanding the flow of alternative fuels.

4.2 LNG Carrier

Natural gas is typically produced in gas fields far from consumers, so it needs to be transported to where it is consumed. For short distances, natural gas is usually compressed into high-pressure gas and transported by pipeline. When the transport distance is very long, or if the shipping destination is an island and deep subsea pipeline is too costly to install, LNG is used. NG is liquefied at a liquefaction plant or LNG-FPSO (floating production, storage, and offloading unit), and LNG carriers transport LNG from the production site to an LNG terminal near the consumption site, as shown in Fig. 4.1.

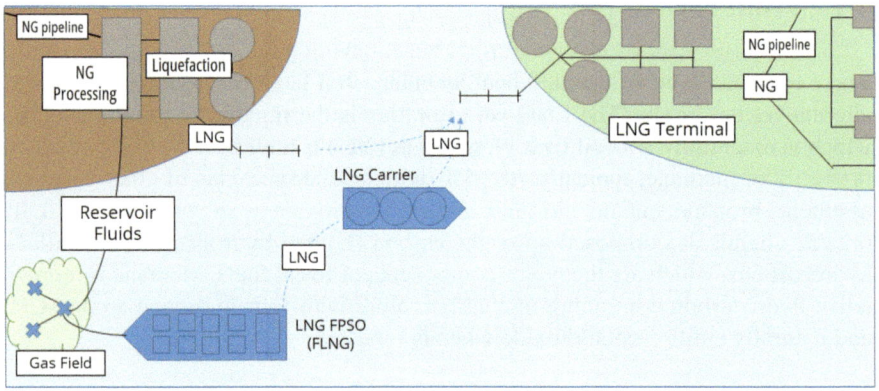

Fig. 4.1 NG supply chain

The main reason for using LNG for long-distance transportation is the difference in density between gas and liquid. For most substances, a liquid phase has a much higher density than the gas phase. Compared to the density of natural gas (about 0.6 kg/m^3), LNG is more than 600 times denser (about 450 kg/m^3) at 1 atm and 25 °C. The increased density significantly increases the mass that can be transported in the same volume of carrier, which improves economics and makes LNG preferable for long-distance transportation. For example, even a carrier with a huge storage tank of 173,000 m^3 can only carry about 100 tons when filled with natural gas at room temperature and pressure. However, when filled with LNG, it becomes possible to transport more than 70,000 tons of LNG.

Methane, the main component of natural gas, has a boiling point of about – 162 °C at atmospheric pressure, which means it needs to be cooled to below – 160 °C to make LNG. Transporting this ultra-cold liquid presents a number of challenges, one of which is the LNG boil-off gas (BOG) problem. Because the temperature of LNG in a storage tank is much lower than room temperature, there is heat ingress from the atmosphere into the tank. Although LNG storage tanks are designed to be thermally insulated to minimize this heat ingress, but even so, it cannot be zero. This leads to the problem of LNG evaporating during transport and forming BOG as shown in Fig. 4.2. When a liquid evaporates in a closed storage tank, it vaporizes and expands in volume, increasing the pressure in the tank. If the pressure exceeds the upper limit of the design pressure that an LNG tank can withstand, an accident such as a natural gas leak can occur. Therefore, it is important to properly handle the BOG before it increases the pressure of the storage tank. Current commercial LNG carriers are designed to be insulated to satisfy boil-off rate (BOR) of 0.1–0.2% per day. While this may seem like a small number, it's not when you consider about the size of a carrier; if you're carrying 70,000 tons of LNG, you're losing at 70–140 tons of evaporated gas per day.

In the past, LNG carriers used steam turbine propulsion to maximize the use of BOG, and the excess BOG even after use was incinerated. However, with the introduction of GHG emission regulations such as EEDI, the use of inefficient steam turbines or incineration of BOG has become a major drawback due to increased GHG emissions, resulting in the development of new engines. With the development of dual-fuel diesel engines (DFDE) and dual-fuel gas injection (GI) engines in the late

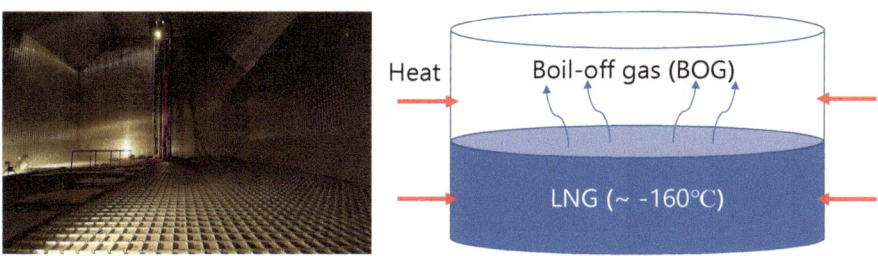

Fig. 4.2 LNG tank and the BOG from LNG in the tank [3]

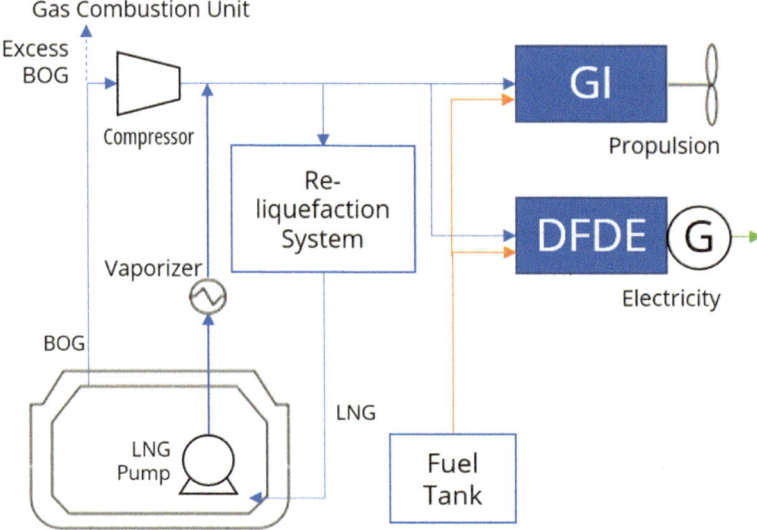

Fig. 4.3 LNG propulsion system having GI, DFDE, and reliquefaction systems

2000s, which can use both natural gas and HFO with high efficiency, LNG carriers are now often designed with a complex system that includes a gas injection engine for propulsion, a DFDE for electricity generation, and a reliquefaction system for excess BOG, as shown in Fig. 4.3.

4.3 LNG Fueled Ship and Bunkering

Technically, an LNG carrier is a ship whose cargo is LNG, not necessarily to use LNG as a fuel for propulsion. However, LNG carriers have commonly used LNG as their primary fuel, as it is often more advantageous to use LNG and BOG from the storage tank.

Since the 2000s, as regulations on SO_x and GHG emissions have tightened, LNG has emerged as an alternative fuel, as it is known to emit almost zero SO_x and lower CO_2 than HFO. In particular, the commercialization of gas injection engines that use NG as a direct fuel has led to the emergence of LNG-fueled ships, which use LNG as the primary propulsion fuel for ships other than LNG carriers.

Figure 4.4a shows the rapidly growing number of ships using alternative fuels, a large proportion of which are LNG-fueled ships. Figure 4.4b shows that the number of LNG-fueled ships is not limited to a specific ship type.

Just as gasoline cars require gas stations, the widespread use of LNG-fueled ships requires the associated infrastructure to deliver LNG to the ships. An LNG fueling

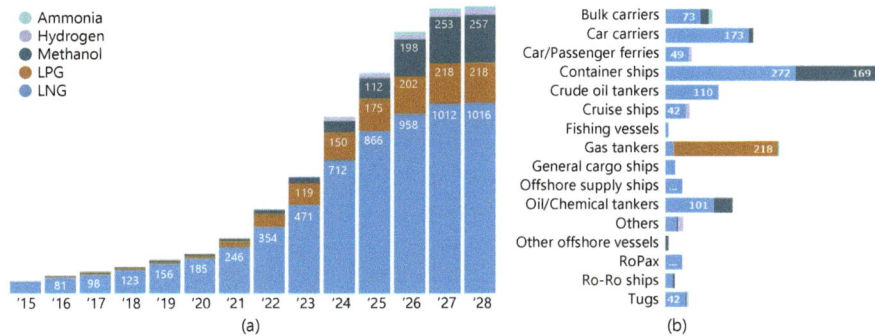

Fig. 4.4   a Ships using alternative fuels by years **b** LNG fueled ship by ships types [1]

vessel is called an LNG bunkering ship. These LNG bunkering facilities are now growing rapidly around the world, as shown in Fig. 4.5.

For LNG bunkering, three methods are commonly used:

1. Supply from onshore LNG storage tanks to ships

   This method requires infrastructure such as a large onshore LNG terminal. Since the main purpose of LNG terminals is to unload LNG from LNG carriers and send it to onshore users, an LNG terminal has both loading and unloading functions. Therefore, by connecting the target ship to the LNG pipes in the terminal, LNG can be delivered directly from the storage tank in LNG terminal on land to the LNG-fueled ship. However, LNG terminals typically have a small number of jetties for large LNG carriers, making it difficult to connect dozens of ships simultaneously. In addition, construction and operating costs are high,

Fig. 4.5   LNG bunkering ships and facilities [1]

so it may be impractical to build new LNG terminals all along the coast in close proximity, like gas stations.

2. Supply from LNG tank lorry to ships

An LNG tank lorry is a specially designed vehicles for storing and transporting liquefied gases. It requires relatively low investment as it can move freely on the road network without the burden of laying pipelines, making LNG available to areas with small demand. However, there is a limit to the amount of LNG that can be stored in a vehicle, so it can be difficult to deliver large volumes of LNG, and bunkering a large ship is not efficient as it may require dozens of vehicles.

3. Supply from LNG bunkering ship to ships

An LNG bunkering ship is a maritime vessel equipped to deliver liquefied natural gas (LNG) to other vessels. It is designed to access ships requiring LNG supply at sea, connect a flexible hose, and supply LNG on the sea. The demand for LNG bunkering ships is increasing because they are not subject to spatial constraints compared to a land-based terminal, and have a larger storage capacity compared to a vehicle.

4.4   Is LNG an Environmentally Friendly Fuel?

LNG is currently thought as a low-carbon emission fuel, but this is mostly based on the carbon dioxide emissions from the combustion phase (TtW). It is now necessary to examine WtW GHG intensity as previously discussed. Figure 4.6a shows the WtW CO_2 emissions intensity of LNG with various LNG engine types (DFMS, DFSS, LBSI) in comparison with conventional marine fuels (HFO, LFO, and MDO/MGO). In the combustion process (TtW), LNG emits less carbon dioxide than conventional fuels. Even after considering the WtW intensity including production, the CO_2 emissions intensity of LNG can be reduced by approximately 18–20% in spite of the type of engine. Therefore, LNG still appears to be a relatively low-carbon emission fuel at first glance.

However, this is based on carbon dioxide intensity only, and the results are different when the GHG intensity is considered. Figure 4.6b shows the CO_2 equivalent GHG emissions intensity of LNG. In the case of a low-speed LNG engine using a diesel cycle (Diesel-DFSS), the GHG intensity is reduced by about 15% compared to HFO, but in the case of a medium-speed LNG engine using an Otto cycle (Otto-DFMS), the GHG emissions intensity is not significantly different from that of HFO even though LNG is used. In other words, even if the same LNG is used as a fuel, the GHG emissions intensity varies depending on which engine is used.

This result is due to the methane slip. Ideally, when a fossil fuel is completely burned, all of the fuel reacts with oxygen to produce only carbon dioxide and water. However, in real engines, incomplete combustion can occur and incomplete combustion in a natural gas engine may cause the unburned methane gas to be released into the atmosphere. This is called as the methane slip. As discussed in Chap. 1, methane

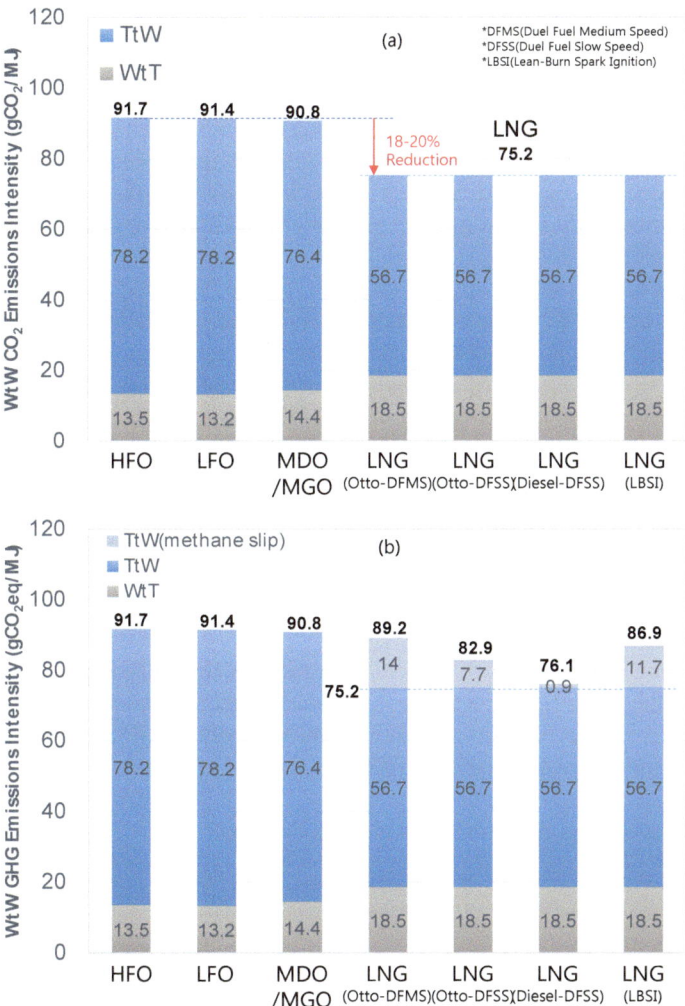

Fig. 4.6 WtW **a** CO_2 and **b** GHG emissions intensities of HFO, LFO, MDO, and LNG based on FuelEU maritime [2]

has a GWP 25–30 times higher than that of carbon dioxide, so even small amounts of methane in the exhaust can have a significant impact.

In other words, it's no longer accurate to say that LNG is an environmentally friendly low-emissions fuel. In Chap. 1, we talked that the important point about alternative fuels is shifting from "what's used" to "how it's produced and used." The same concept applies to LNG. The GHG emissions intensity of LNG can be significantly different depending on how it is produced and used.

4.5 Environmentally Friendly LNG

However, it is premature to say that LNG is already an obsolete fuel. As we will discuss in Chap. 5 on hydrogen, the demand for natural gas as a feedstock would be maintained because it is currently the primary feedstock for alternative fuels such as hydrogen. In addition, the lifetime of LNG as a fuel could be extended in combination with carbon capture, utilization, and storage (CCUS) technologies discussed in Chap. 7.

In addition, environmentally friendly LNG can be produced from some resources, such as biogas. Biogas is a gas produced by anaerobic fermentation that decomposes organic components contained in sewage, agricultural waste, food waste, or municipal solid waste (MSW) as shown in Fig. 4.7. The main components of biogas are mostly methane and carbon dioxide, and the methane content is usually more than 50%. By separating carbon dioxide and other impurities from biogas, biomethane based on non-fossil fuel can be obtained. After liquefying it, bio-LNG can be produced. As will be discussed in Chap. 6, a non-fossil based biofuel could be a sustainable alternative fuel if it can significantly reduce the net GHG emissions. Currently, biogas or biomethane is still mostly used as a heat source near the source of the gas, but there is a growing movement around the world to use the biogas and biomethane efficiently.

In summary, while the use of LNG as a fuel is expected to decline gradually in the future, the use of LNG as a feedstock is expected to continue for a considerable period of time. Moreover, even as the use of NG from fossil fuels declines, the use of biomethane based on non-fossil fuels is expected to continue.

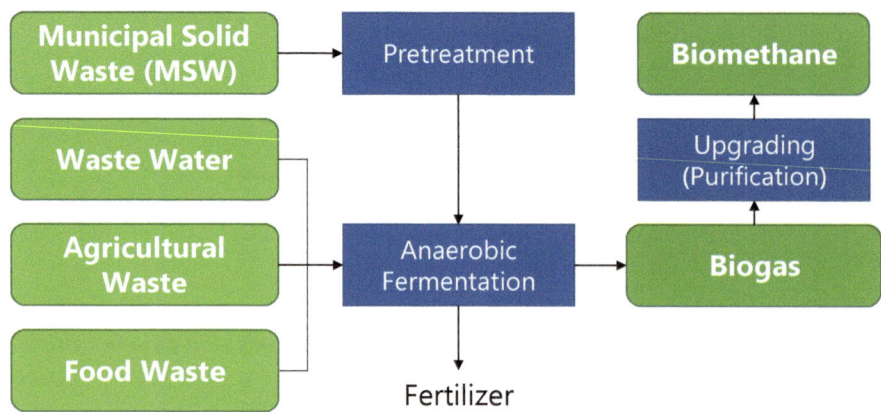

Fig. 4.7 Biogas and biomethane production process

References

1. DNV AFI, Alternative Fuels Insight (2024). https://afi.dnv.com/statistics/. Accessed 30 Oct 2024
2. Regulation (EU) 2023/1805 of the European Parliament and of the Council of 13 September 2023 on the use of renewable and low-carbon fuels in maritime transport. https://eur-lex.europa.eu/eli/reg/2023/1805/oj
3. Thinfourth, Mark III type LNG membrane inside an LNG carrier (2010). https://en.m.wikipedia.org/wiki/File:Liquid_natural_gas_membrane_tank.jpg. Accessed 30 Oct 2024

Chapter 5
Hydrogen and Ammonia

Abstract Hydrogen and ammonia are gaining attention as alternative fuels due to
their carbon-free combustion feature, and offer a potential solution for decarbonization. However, they also face challenges to overcome. Currently, most hydrogen
is produced from natural gas through steam methane reforming (SMR), which
produces significant GHGs, called as "gray" hydrogen. In contrast, "green" hydrogen,
produced by water electrolysis using renewable energy, generate little GHGs but not
yet economically feasible due to high production costs. Liquefied hydrogen (LH_2)
has logistical challenges, such as low energy density and high liquefaction costs, as
well as the risk of GHG emissions from the energy-intensive liquefaction process.
Ammonia is easier to transport due to its milder liquefaction conditions, and can be
used directly as a fuel once the ammonia-fueled engines currently under development
are commercialized. However, like hydrogen, the environmental impact of ammonia
depends on how it is produced. "Gray" ammonia, synthesized from fossil fuels,
has high GHG emissions, while green ammonia, produced from green hydrogen,
is more sustainable but still requires careful consideration of its energy-intensive
synthesis process. Additionally, ammonia is a toxic substance that poses health risks
and requires stringent safety measures.

Keywords Hydrogen · Ammonia · Color classification · Gray hydrogen · Green
hydrogen · Gray ammonia · Green ammonia

5.1 Why Hydrogen Is of Interest?

Fossil fuels such as coal, oil, and natural gas, which are the primary fuels that humans
have used, are hydrocarbon mixtures. Hydrocarbons are based on carbon, and as a
result, the formation of carbon dioxide after combustion is inevitable. Hydrogen has
recently received a lot of attention because it is a molecule (H_2) with two hydrogen
atoms (H), and when completely combusted with oxygen (O_2), only water (H_2O) is
produced without the formation of carbon dioxide.

Y. Lim, *Alternative Fuels for Environmentally-Friendly Ships*,
SpringerBriefs in Applied Sciences and Technology,
https://doi.org/10.1007/978-3-031-85082-0_5

$$H_2 + 0.5O_2 \rightarrow H_2O$$

Furthermore, research about hydrogen-fueled propulsion has a quite long history. Space launch vehicles with liquid hydrogen propulsion systems were already used as early as the 1960s (Fig. 5.1), and BMW produced and sold 100 hydrogen-fueled vehicles called Hydrogen 7 in 2005–2007. More recently, hydrogen fuel cell electric vehicles (FCEVs) with hydrogen fuel cells have been commercialized, sold, and operated.

Fig. 5.1 Atlas-Centaur launching Surveyor 1 using liquid hydrogen as propellants [5]

5.2   Is Hydrogen an Environmentally Friendly Fuel?

Since Chap. 1, it has been noted for alternative fuels that we should focus on how the fuel is produced, not what the fuel is. The reason for this is very clear in the case of hydrogen. Hydrogen is a zero carbon emission fuel if only the combustion stage is evaluated, as no GHGs are produced during the combustion stage. However, as shown in Fig. 5.2, the WtW GHG emissions intensity of hydrogen is 40% higher than that of HFO, when it is reformed from natural gas (referred to as gray hydrogen). This is because excessive GHGs are emitted during the hydrogen production stage (WtT), even though no GHGs are generated during the combustion stage (TtW). In other words, hydrogen produced from natural gas emits more GHGs than petroleum, so it is clear that it is not the sustainable alternative fuel that people expect.

There are two main reasons why the WtW GHG emissions intensity of gray hydrogen is high. Figure 5.3 shows a schematic diagram of the steam methane reforming (SMR) process, a conventional hydrogen production process in widespread commercial use. There are two main reactions used to produce hydrogen: the SMR reaction, in which methane reacts with steam to produce synthesis gas, a mixture of carbon monoxide (CO) and hydrogen, and the water gas shift (WGS) reaction, in which carbon monoxide and water react to produce hydrogen. This means that the hydrogen produced by reforming natural gas in this way inevitably generates carbon

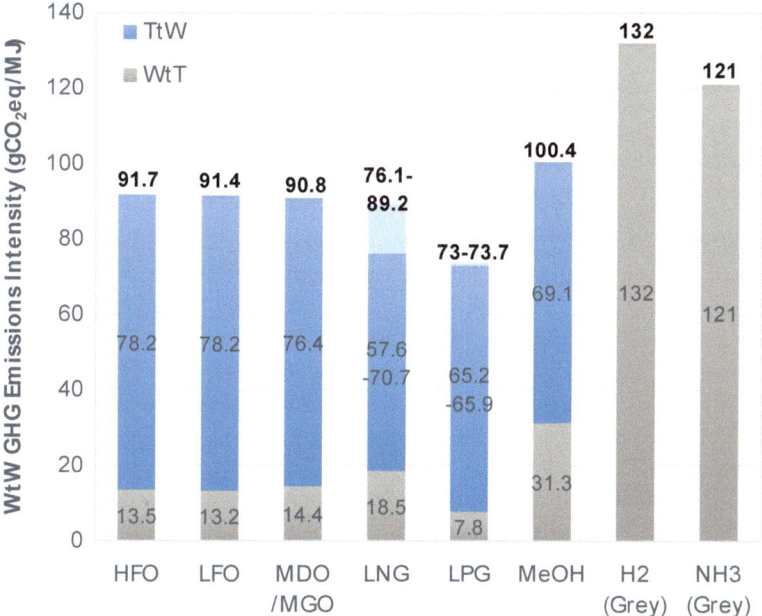

Fig. 5.2 WtW GHG emissions intensity of various fuels based on FuelEU maritime [6]

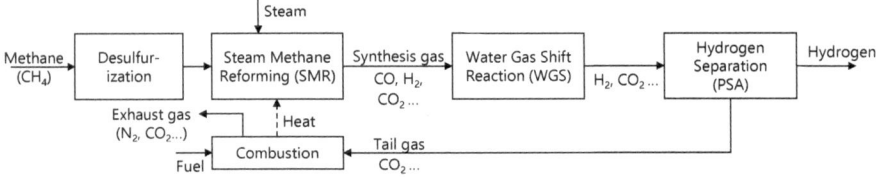

Fig. 5.3 Steam methane reforming (SMR) process for representative hydrogen production

dioxide during the production process, and when it is emitted, it emits carbon dioxide that originates from the natural gas, which is a fossil fuel.

$$\text{SMR reaction: } CH_4 + H_2O \rightleftharpoons CO + 3H_2$$
$$\text{WGS reaction: } CO + H_2O \rightleftharpoons CO_2 + H_2$$
$$\text{Overall reaction: } CH_4 + 2H_2O \rightleftharpoons CO_2 + 4H_2$$

Second, the SMR reaction requires high temperatures of 700–900 °C and high pressures of 5–25 bar to achieve sufficient conversion rates. To make reaction conditions of high temperature and pressure, the fluid must be heated or compressed, which in turn requires energy. If this energy is supplied from fossil fuels, additional greenhouse gases are generated.

5.3   Color Classification of Hydrogen

Hydrogen can be classified by color as shown in Fig. 5.4 to indicate how it was produced and the energy sources used.

- Black/Brown hydrogen: Hydrogen produced from coal gasification.
- Gray hydrogen: Hydrogen obtained by steam reforming fossil fuels like natural gas.
- Green hydrogen: Hydrogen produced by electrolyzing water, using renewable energy sources like solar or wind which does not generate GHGs.
- Blue hydrogen: Hydrogen produced from fossil fuels such as natural gas, but CO_2 emissions are reduced by using carbon capture, utilization and storage (CCUS) technologies.
- Pink hydrogen: Hydrogen produced by nuclear power as the primary energy source.

The environmentally friendly hydrogen that people usually expect is the green hydrogen, which emits little or no GHGs. However, currently green hydrogen is not easy to use in our daily lives because of its high unit production cost. Figure 5.5 shows the levelized cost of hydrogen production estimated by the IEA, which shows that green hydrogen produced from water electrolysis is more than 2–4 times more

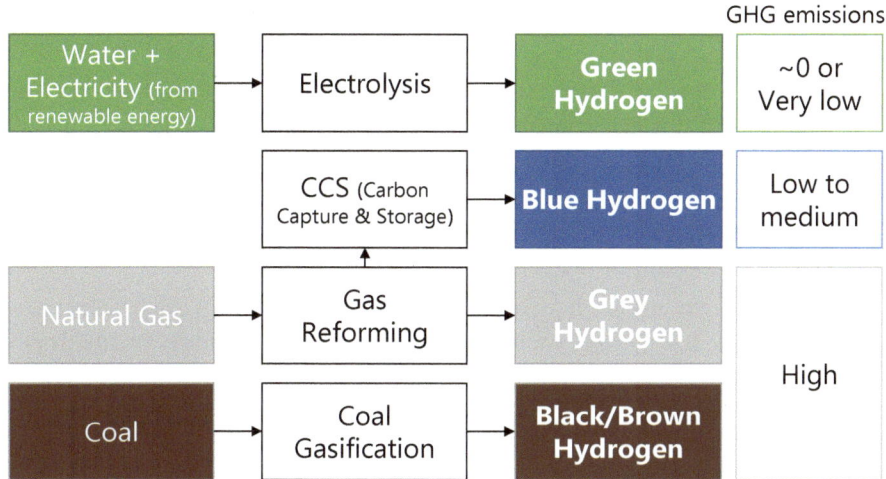

Fig. 5.4 Color classification of hydrogen

expensive than gray hydrogen produced from natural gas. The main reason is not only the investment cost of water electrolysis, but also the cost of electricity generation. Even if the investment cost of water electrolysis technology could be reduced, it is inevitable that high operating costs will be required if high-cost electricity is used to provide the electricity required for water electrolysis.

In addition, even if hydrogen is produced by water electrolysis, it cannot be green hydrogen if the electricity is generated by fossil fuels such as coal-fired power plants. In other words, the following three conditions must be met for green hydrogen to be used. First, the electricity required for water electrolysis must be generated from

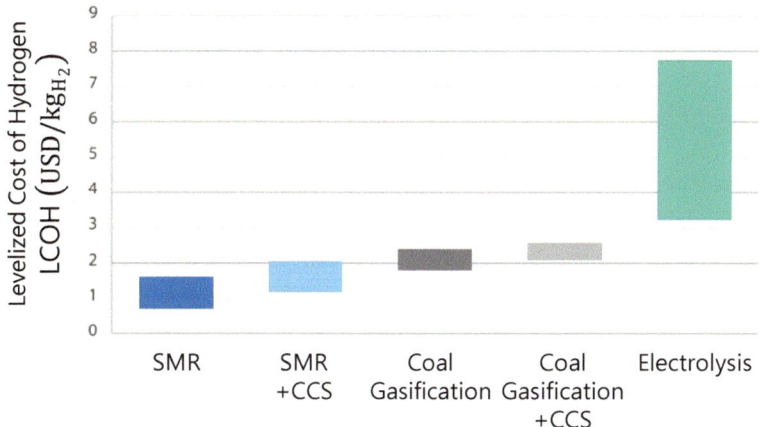

Fig. 5.5 Levelized cost of hydrogen production [3]

renewable energy sources such as wind and wave power which emit near-zero carbon dioxide. Second, the cost of renewable electricity must be low enough. Third, the investment and operating costs of water electrolysis must be low enough. For these reasons, green hydrogen is not economically feasible right now.

5.4 Liquefied Hydrogen (LH$_2$)

For long distance transport, liquefied hydrogen transport by ship can be considered, similar to LNG. A challenge for hydrogen transportation is its low energy density. Energy density is the amount of fuel required to produce a unit energy, and Fig. 5.6 shows the energy density of hydrogen compared to methane. Compressed hydrogen, even when compressed to high pressures of 350 or 700 bar, has only 14–24% of the energy density of liquefied methane. Liquefied hydrogen has a higher energy density than high pressure compressed hydrogen, but it is still only 43% of liquid methane. This means that the volume required to transport hydrogen will be much higher than for LNG.

Hydrogen liquefaction technology itself has been commercially available since the early 1900s, and, but it is a costly and energy-intensive process because the temperature required to liquefy hydrogen at atmospheric pressure is – 253 °C, which is much lower than LNG, which requires temperatures around – 160 °C. Figure 5.7 shows the specific energy consumption of the hydrogen liquefaction process, which is currently around 12–14 kWh/kg for commercially available hydrogen liquefaction processes. Considering that the specific energy consumption for LNG is around 0.4–1.0 kWh/kg, hydrogen liquefaction requires a very high level of energy. Some studies have shown that the specific energy consumption for hydrogen liquefaction can be reduced to 5–9 kWh/kg, but this has not yet been commercially demonstrated.

The high energy consumption is not only a cost issue, but also increases GHG emissions depending on where the energy comes from. If green hydrogen is produced

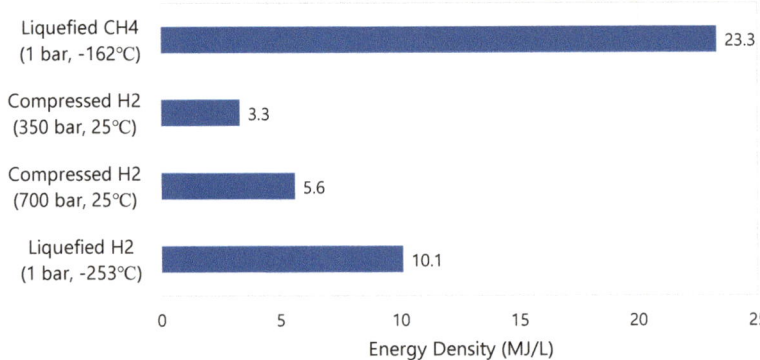

Fig. 5.6 Energy density of methane and hydrogen

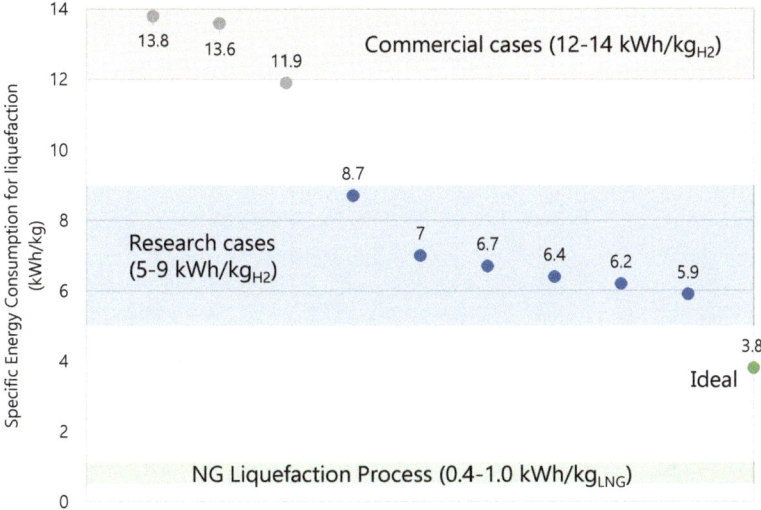

Fig. 5.7 Specific energy consumption for hydrogen liquefaction ([4]; data from [1])

without GHG emissions, but it is liquefied by using a petroleum-based energy, the GHG emissions from the liquefaction process will end up being included in the WtW GHG emissions of the hydrogen. Then, the hydrogen may not be green. The discussion at the 80th MEPC on WtW GHG emissions was important because of these complex issues.

There is another problem transporting liquefied hydrogen by ship. It's the same problem as with LNG: boil-off gas (BOG). The temperature of liquefied hydrogen is very low, − 253 °C at atmospheric pressure, which is 90 °C lower than LNG, so more heat enters from the atmosphere. In addition, hydrogen has a lower heat of vaporization than methane, resulting in a higher boil-off rate (BOR). Therefore, to maintain the same BOR with LNG, higher insulation performance is required for liquid hydrogen.

Currently vacuum insulation technology is applied to liquid hydrogen tank, and various advanced insulation methods are currently being researched for large-scale liquid hydrogen transportation. In Japan, Kawasaki Heavy Industries demonstrated the operation of a liquid hydrogen carrier with two 1250 m³ LH₂ storage tanks with vacuum perlite insulation in 2020 (Fig. 5.8), and the conceptual design of a 160,000 m³ LH₂ carrier with four 40,000 m³ tanks was certified in 2022. South Korean shipbuilders such as HD Hyundai Heavy Industries, Samsung Heavy Industries, and Hanhwa Ocean have also received design approval for LH₂ carriers.

Fig. 5.8 Suiso Frontier at Kawasaki Heavy Industries Kobe Shipyard [2]

5.5   Why Ammonia is of Interest?

The interest in ammonia as an alternative fuel is linked to the aforementioned shortcomings of liquefied hydrogen transport. If transporting liquefied hydrogen is energy-intensive and costly, why not turn it into a substance more easily transported? Ammonia (NH_3) is used as a raw material for a variety of compounds, including fertilizers, and has been commercially produced in large quantities since the early 1900s using the Haber–Bosch process as shown in Fig. 5.9, which involves the synthesis of hydrogen (H_2) and nitrogen (N_2).

$$N_2 + 3H_2 \rightleftharpoons 2NH_3$$

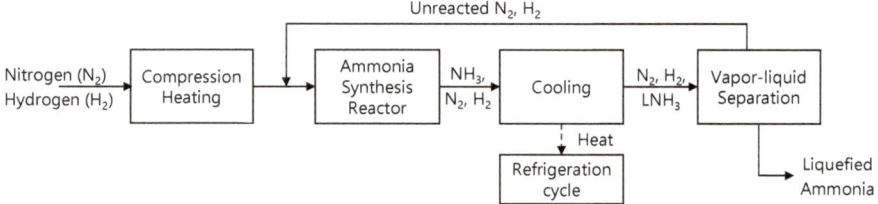

Fig. 5.9 Conventional Haber–Bosch process for ammonia synthesis

Ammonia does not contain carbon, so it does not produce carbon dioxide when combusted. Ammonia can be liquefied at about − 33 °C at atmospheric pressure, or about 10 bar at 25 °C. These are much milder conditions compared to the storage conditions for liquefied hydrogen, and similar to the LPG storage condition which is commercially available. In addition, ammonia dual-fuel engines, which can directly use ammonia as a fuel are currently being developed by various engine companies such as HD Hyundai, WinGD, and MAN ES.

5.6 Is Ammonia an Environmentally Friendly Fuel?

Figure 5.2 shows that gray ammonia, which is synthesized from gray hydrogen, has a 30% higher WtW GHG emissions intensity than that of HFO. In other words, just like hydrogen, ammonia cannot be considered as an environmentally friendly alternative fuel if it is produced from fossil fuels. Extending the color classification used for hydrogen, ammonia can also be classified into gray, blue, and green ammonia, as shown in Fig. 5.10. Not surprisingly, green ammonia from green hydrogen is closer to the green alternative fuel that people hope for.

Let's assume that we have produced an ideal green hydrogen with zero GHG emissions from renewable energy, and we synthesized ammonia from it. Does this ammonia have zero WtW GHG emissions intensity? Probably not. The conventional Haber–Bosch process used to synthesize ammonia requires reaction conditions at high temperatures of 300–500 °C and high pressures of 100–200 bar in the presence of a suitable catalyst. Making high temperatures and pressures requires energy, and if that energy comes from fossil fuels, it produces GHGs. To separate the ammonia from unreacted hydrogen and nitrogen, the gas is usually cooled through a refrigeration cycle to liquefy and separate the ammonia. The refrigeration cycle also requires energy and it may emit GHGs. In addition, the ammonia synthesis process requires nitrogen as well as hydrogen. Conventionally nitrogen is separated from air by consuming energy, and it may emit GHGs. In the end, green ammonia cannot be

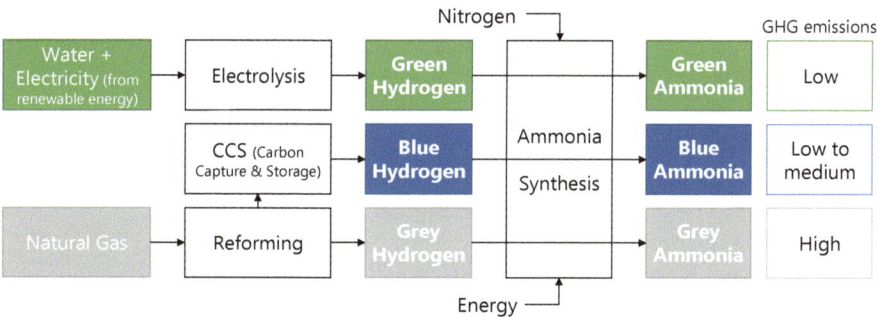

Fig. 5.10 Color classification of ammonia

Table 5.1 Ammonia hazard

Concentration (ppm)	Hazard to human health
5	Unique detectable smell
20	Eye irritation and problems in the respiratory system
200	Headaches and irritation of nose and throat
700	Eye damage
1700	Coughing and difficulty of breathing
4500	Fatal damage
5000+	Possible death due to respiratory arrest

produced from green hydrogen alone, but requires a comprehensive analysis of the production process.

There is an additional issue to consider when using ammonia as a fuel. It was mentioned in Sect. 4.4 that in a gas injection engine using NG as a fuel the methane slip may occur, where incompletely combusted methane is released into the exhaust. Similarly, in an engine using ammonia as a fuel, the ammonia slip may occur, where incompletely combusted ammonia is released into the exhaust. The serious problem is that ammonia is a toxic substance. At very low concentrations, ammonia causes only a foul odor, but as concentrations increase, it may even cause casualties, as shown in Table 5.1. Most countries have set standards for acceptable concentrations of ammonia in the workplace. The permissible exposure limit is given as a time weight average concentration (TWA), which are based on an 8-h working day, or a short-term exposure limit (STEL), which are based on 15 min of contact; for ammonia, the TWA is 25 parts per million (ppm) and the STEL is 35 ppm. Therefore, for the ammonia engines under development, ammonia slip or leakage should be minimized and a safety system is required to deal with any leakage.

Hydrogen and ammonia, with their carbon-free molecular structure, are seen as the next generation of alternative fuels for the decarbonization era. However, a closer look reveals that there are still many challenges to overcome. It is important to recognize that the simple use of hydrogen or ammonia cannot be an environmentally friendly solution. How it is produced and used is a critical issue that must be included in any future discussion of environmentally friendly alternative fuels.

References

1. S.Z.S. Al Ghafri et al., Hydrogen liquefaction: a review of the fundamental physics, engineering practice and future opportunities. Energy Environ. Sci. (2022)
2. Hunini, Suiso Frontier (2020). https://commons.wikimedia.org/wiki/File:SUISO_FRONTIER_left_rear_view_at_Kawasaki_Heavy_Industries_Kobe_Shipyard_October_18,_2020_03.jpg. Accessed 30 Oct 2024
3. IEA, Projected Costs of Generating Electricity 2020 (2020). https://www.iea.org/reports/projected-costs-of-generating-electricity-2020

4. Y. Lim, *Understanding Environmentally Friendly Ships* (Sungandang, Seoul, 2023)
5. NASA, Great Images in NASA GPN-2000-000617 (1966). https://commons.wikimedia.org/wiki/File:Surveyor_1_launch.jpg
6. Regulation (EU) 2023/1805 of the European Parliament and of the Council of 13 September 2023 on the use of renewable and low-carbon fuels in maritime transport. https://eur-lex.europa.eu/eli/reg/2023/1805/oj

Chapter 6
Biofuel and E-fuel

Abstract Biofuels such as biodiesel, bioethanol, biomethanol, and biomethane are fuels produced from biomass, which includes organic matter. While biofuels reduce net GHG emissions by absorbing CO_2 during their growth, their overall impact depends on the balance between CO_2 absorbed and emitted during production and combustion. First-generation biofuels from edible crops have a problem of competition with food and may lead to higher food prices. Additionally, the effects of land use change (LUC), such as deforestation, can exacerbate GHG emissions. As a result, the EU and IMO have developed guidelines for biofuel criteria, focusing on WtW GHG emissions intensity from production to use of biofuels. In the future, advanced biofuels produced from non-edible biomass, may offer a more sustainable alternative. E-fuels produced from renewable hydrogen and captured CO_2 are also gaining attention, as they can reduce WtW GHG emissions and be used without modifying conventional fossil fuel-based systems. However, the benefits of e-fuels depend on the source of CO_2, because using of non-renewable CO_2 from fossil fuels may not reduce the WtW GHG emissions sufficiently. While challenges remain, both biofuels and e-fuels are seen as key options for achieving net-zero emissions in the future.

Keywords Biofuel · Biodiesel · Bioethanol · Biomethanol · Advanced biofuel · E-fuel · Renewable CO_2 · Non-renewable CO_2

6.1 What is a Biofuel?

Biofuels are fuels produced from biomass, which refers to organic matter that can be converted into energy. Biomass includes all living organisms, including plants, which synthesize energy from water and carbon dioxide through photosynthesis, and the animals and micro-organisms that feed on them. Since biofuels contain carbon, it is inevitable that GHGs such as carbon dioxide are emitted ($+\alpha$) when they are combusted. However, biomass absorbs carbon dioxide through photosynthesis as it grows, so GHG emissions to the atmosphere are reduced ($-\beta$). Therefore, net GHG emissions become the difference between GHG absorption and emissions ($\alpha-\beta$), so

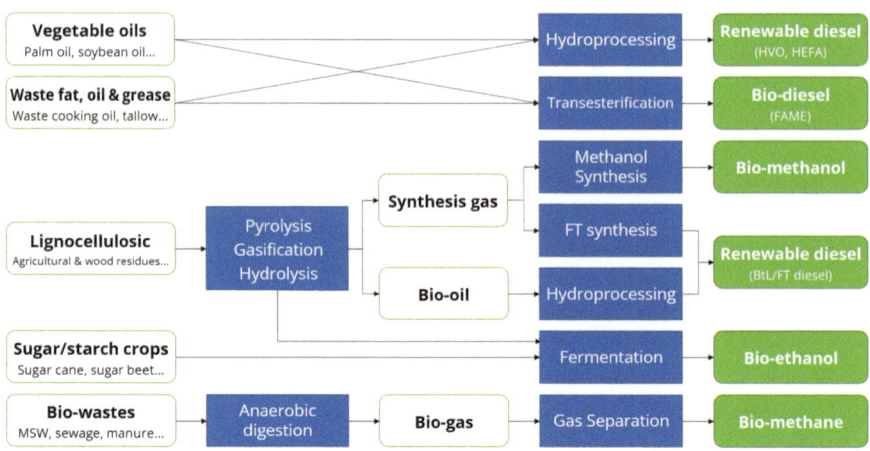

Fig. 6.1 Biofuels from various biomasses and processes

a biofuel can be considered as an alternative fuel if the net GHG emissions are low enough.

Currently various biofuels can be produced from a wide range of biomasses, such as vegetable oils, waste oils, straw or waste wood, municipal solid waste (MSW), or sewage, as shown in Fig. 6.1. Biofuel products include biomethane, biomethanol, bioethanol, biodiesel, and even alternative marine fuels for ships, such as bio marine oil based on biodiesel. In other words, today's petroleum-based fuel systems can be used without major changes if biofuels can be ideally produced.

6.2   Biodiesel, Marine Biofuel, and Biomethanol

Biodiesel is a mixture of fatty acid methyl esters (FAMEs), which can be produced by reacting triglycerides (the main component of animal and vegetable fats/oils) with alcohols in the presence of a catalyst as shown in Fig. 6.2. Because FAME blends have similar combustion properties to diesel, they are called as biodiesel. Depending on its composition, biodiesel has poor fluidity at low temperatures and high freezing point, so blends of FAME with diesel such as BD10 (10% biodiesel with 90% diesel), BD20 (20% biodiesel with 80% diesel), etc. are commonly used.

The fuel blended with diesel, biodiesel, and biodiesel by-products to meet the quality of marine fuel oil is called marine biofuel or bio marine fuel. Marine biofuel has already been demonstrated. ExxonMobil, one of the international oil majors, tested marine biofuel with the international shipping company Stena Bulk in 2020, and the Korean shipping company HMM successfully applied marine biofuel to a container ship operating from Busan to the Panama Canal in 2021.

When biomass, such as waste wood, is indirectly heated in the absence of an oxidizing agent such as oxygen, the components of the biomass decompose to

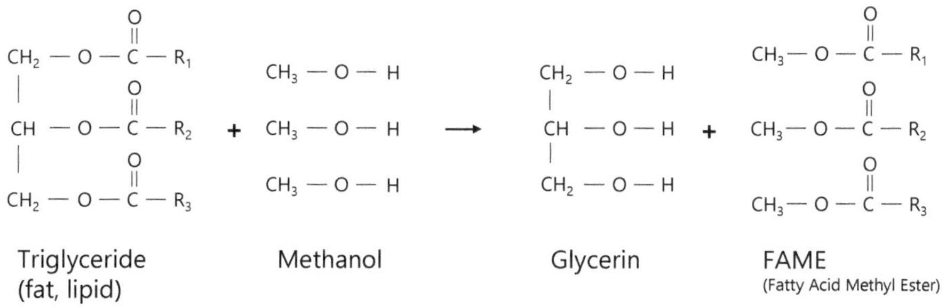

Fig. 6.2 FAME (biodiesel) formation from triglyceride

produce synthesis gas (syngas), a mixture of hydrogen and carbon monoxide. In the presence of a suitable catalyst, biomethanol can be synthesized from the syngas.

$$CO + 2H_2 \rightleftharpoons CH_3OH$$

Methanol can be used as a marine fuel. Man ES and HD HHI demonstrated and commercialized marine methanol propulsion engines in 2022. The world's largest shipping company, AP Moller–Maersk has announced that it will use green methanol from 2023.

6.3 Diesel Versus Doughnuts Debate

So if biofuels are environmentally friendly fuels, shouldn't we just switch to biofuels for all of fuels to reduce GHG emissions? That's not quite simple, because there are still a number of challenges. Most of the biofuels currently being produced are first-generation biofuels, which are based on edible grains or oils. In 2021, the Financial Times reported that the price of soybean oil has skyrocketed as US refiners use it as a feedstock for biodiesel, which in turn has increased the price of doughnuts. This is known as the "diesel vs doughnuts debate" [7].

The feedstocks for first-generation biofuels are mainly edible resources such as soybean or palm oil. There is a limited supply of edible crops, so when a supportive policy is introduced to increase the production of a biofuel, the price of its feedstock rises as food and biofuel producers compete for the feedstock. In fact, in 2007, the price of corn, the main feedstock for bioethanol, temporarily rose sharply when the US Congress mandated the blending of bioethanol into gasoline.

A shortage of edible crops not only increases the unit cost of a particular food crop, but it can also raise the price of all other processed foods made from it. Corn, for example, is not only consumed directly as human food, but it is also used as an ingredient in secondary processed food products such as corn syrup (liquid fructose), and as feed for livestock such as pigs, cows, cattle, and so on. This means that when the price of corn goes up, not only does the price of edible corn goes up, but the price of processed foods made from corn, beverages, beef, milk, etc., all go up.

Therefore, biofuels derived from edible crops and their by-products can lead to a combination of food crop scarcity and rising food prices. This ties in with the food security debate about whether to prioritize food or biofuels, which can be a very sensitive issue, especially in countries where food costs are a high proportion of household income/expense, or where countries are dependent on food imports.

6.4   Land Use Change (LUC)

Another debate surrounding biofuels is the land use change (LUC). In order to produce crops for biofuels, arable land is needed to grow the crops. Cultivation of land leads to direct and indirect changes in land use, which are closely linked to greenhouse gas emissions. For example, burning forested areas to make cropland results in the loss of trees and soil biological organisms that hold carbon dioxide, thereby increasing the amount of GHGs in the atmosphere and destroying the forest's ability to absorb carbon dioxide. It has therefore been consistently argued that biofuel production should consider the impact from LUC.

LUC is categorized in two ways: direct land use change (dLUC) and indirect land use change (iLUC). For example, if farmer A switches from growing wheat to growing sugar cane to produce biofuel feedstock, this is a direct land use change (dLUC). If A's LUC leads to a shortage of wheat flour and an increase in the price of wheat causes that another farmer, B, to convert grassland to wheat field, this is an indirect land use change (iLUC). The GHG emissions from biofuels can increase significantly if these LUCs are included. In particular, many researchers have reported that converting savannah grassland or tropical rainforest to cropland can lead to a significant increase in GHG emissions as shown in Fig. 6.3.

Estimating the impacts of direct and indirect land use change was a challenge, because the estimation methodologies differed between researchers, and the resulting GHG emissions were not the same. However, as the importance of LUC continues to be highlighted and specific methodologies have been established by various researchers, the inclusion of LUC impacts in the GHG emissions of biofuels has been gradually adopted in recent years. The International Organization for Standardization (ISO) published "ISO 14067:2013 Greenhouse gases—Carbon footprint of products—Requirements and guidelines for quantification and communication" in 2013 and revised it in 2018 [3]. According to this document, dLUC is a factor that shall be considered when calculating the carbon footprint, and consideration of iLUC is also recommended.

The European Union (EU) is preparing various regulations and directives to reduce GHG emissions under the "Fit for 55" policy, which aims to reduce greenhouse gas emissions by at least 55% by 2030 compared to 1990 levels. As part of this policy, the EU adopted the EU Renewable Energy Directive (RED) in 2009 and revised RED II in 2018 to introduce standards to reduce GHG emissions across the EU. In

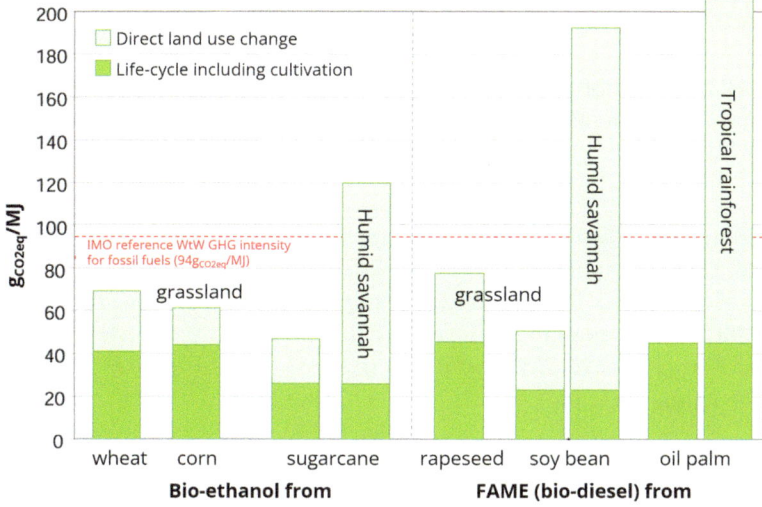

Fig. 6.3 Life-cycle GHG emissions of biofuels and impacts from direct LUC (excluding indirect LUC) [4]

particular, it sets sustainability criteria for biofuels, limiting the use of edible food biomass (wheat, corn, etc.), which has a high potential to cause iLUC, to a maximum of 7% across the EU and phasing it out by 2030. RED II also introduced provisions to promote the advanced biofuels, which are based on non-edible food biomass such as straw, agricultural by-products, wood by-products, and algae. Examples of advanced biofuel feedstocks currently covered by RED II are as follows [1].

- Algae if cultivated on land in ponds or photobioreactors
- Biomass fraction of mixed municipal waste (not separated household waste for recycling)
- Biomass fraction of industrial waste not fit for use in the food or feed chain
- Straw
- Animal manure and sewage sludge
- Palm oil mill effluent and empty palm fruit bunches
- Tall oil pitch
- Crude glycerine
- Bagasse
- Grape marcs and wine lees
- Nut shells
- Husks
- Cobs cleaned of kernels of corn
- Biomass fraction of wastes and residues from forestry and forest-based industries
- Non-food cellulosic material
- Ligno-cellulosic material except saw logs and veneer logs
- Used cooking oil
- Animal fats.

6.5 IMO Interim Guidance on the Use of Biofuels

At the 80th session of the IMO MEPC in 2023, the "interim guidance on the use of biofuels" was adopted. The two key points of this guidance are summarized as follows [5].

> **Pending the development of the comprehensive method to account for WtW GHG emissions and removals based on the LCA Guidelines [6], biofuels that have been certified by an international certification scheme, meeting its sustainability criteria, and that provide a WtW GHG emissions reduction of at least 65% compared to the WtW emissions of fossil MGO of 94 g_{CO_2eq}/MJ, may be assigned a C_f equal to the value of the WtW GHG emissions of the fuel according to the certificate multiplied by its lower calorific value.**

> **Biofuels not certified as "sustainable" or not fulfilling the WtW emission factor criterion above should be assigned a C_f equal to the C_f of the equivalent fossil fuel type.**

This means that even if a biofuel is produced from biomass, it will be treated as a fossil fuel if its WtW GHG reduction cannot meet the 65% reduction requirement (i.e. achieving an intensity not exceeding 33 g_{CO_2eq}/MJ). The IMO's methodology for estimating the GHG emissions intensity of biofuels is not yet been finalized and is therefore subject to change. However, for example, estimates based on EU RED II and the FuelEU maritime initiative show that biofuels from edible biomass feedstocks are unlikely to meet the 65% reduction requirement as shown in Fig. 6.4. Furthermore, it is expected that in the future the reduction targets will be gradually increased from the current 65% reduction.

This shows that the way we look at biofuels, like other alternative fuels, is shifting from the "what" to the "how." That is, it's not important whether a biofuel is used, but how it is made from what feedstock. The criteria for qualifying a biofuel as an environmentally friendly alternative fuel are also moving toward a quantitative assessment of GHG emissions reductions.

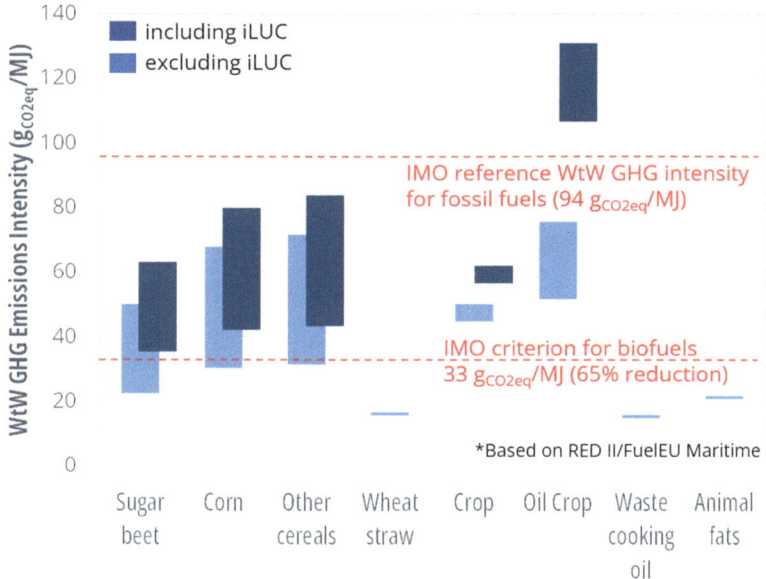

Fig. 6.4 WtW GHG emissions intensity of biofuel by biomass feedstock

6.6 FT Process and Synthetic Fuel

Since the twentieth century, various studies have been conducted to synthesize liquid fuels from coal or natural gas, that can replace petroleum. The technologies for producing liquid synthetic fuels from gas are collectively referred to as gas-to-liquid (GTL) technologies, and the Fischer–Tropsch synthesis process is a representative technology. The Fischer–Tropsch synthesis process, or FT process for short, is a process for synthesizing a mixture of liquid hydrocarbons that can replace petroleum by reacting synthesis gas (syngas), a mixture of carbon monoxide (CO) and hydrogen (H_2), at high temperatures and pressures (conventionally around 200–350 °C and 10–30 bar) in the presence of a suitable metal catalyst, such as iron or cobalt compounds. It was first developed by German scientists Fischer and Tropsch in the 1920s, and has been improved upon many times since. The main reactions of FT process is as follows.

$$(2n + 1)H_2 + nCO \rightleftharpoons C_nH_{2n+2} + nH_2O$$

where n is an arbitrary integer, different compounds are produced depending on the reaction conditions. For example, when $n = 7$ and $n = 8$, heptane and octane are produced, respectively, which are the main components of gasoline. The FT process has been demonstrated and operated commercially for decades, so there is a high level of technical maturity.

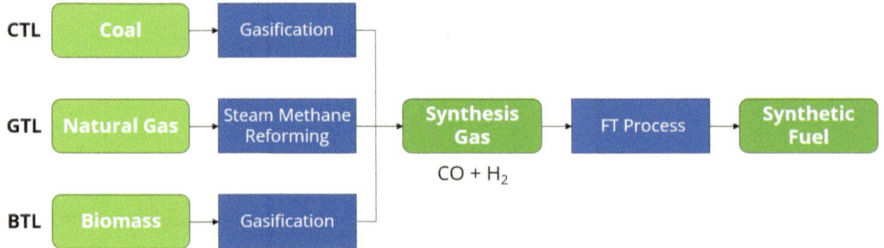

Fig. 6.5 Synthetic fuel produced from different feedstocks using the FT process

$$n = 7: 15H_2 + 7CO \rightleftharpoons C_7H_{16} + 7H_2O$$
$$n = 8: 18H_2 + 8CO \rightleftharpoons C_8H_{18} + 8H_2O$$

Syngas can be produced from a variety of feedstocks. When natural gas is used as a feedstock, reactions are used to convert methane into syngas. A typical example is the steam methane reforming (SMR) reaction, which is used to produce hydrogen.

$$CH_4 + H_2O \rightleftharpoons CO + 3H_2$$

The FT process can also be used with coal gasification technology to produce liquid synthetic fuels from coal, which is referred to as coal-to-liquid (CTL) technologies. When syngas is produced from biomass to synthesize liquid fuel, the process is called as biomass-to-liquid (BTL) technologies. In other words, there are a variety of technologies that can produce liquid synthetic fuels from a variety of resources, including coal, natural gas, and biomass as shown in Fig. 6.5.

6.7   What is an E-fuel?

As we've seen in Sect. 6.6, not all synthetic fuels are environmentally friendly, because they can be produced from fossil fuels such as coal or natural gas. If only biomass is used as a feedstock, synthetic biofuels can be produced, which partially reduces greenhouse gas emissions as discussed in Sect. 6.1.

However, if we look at the basic principle of the FT process, we can know that a liquid fuel can be synthesized using hydrogen and carbon monoxide as raw materials. So what if we generate electricity from renewable energy and electrolyze water to produce green hydrogen, and then produce carbon monoxide from carbon dioxide captured from the atmosphere, and then synthesize them to make synthetic fuels?

Under the proper reaction conditions, water can react with CO to form carbon dioxide and hydrogen. This is called the water gas shift (WGS) reaction, which was discussed in hydrogen production. By changing the reaction conditions, this reaction can be reversed, which is called the reverse water gas shift (RWGS) reaction, meaning

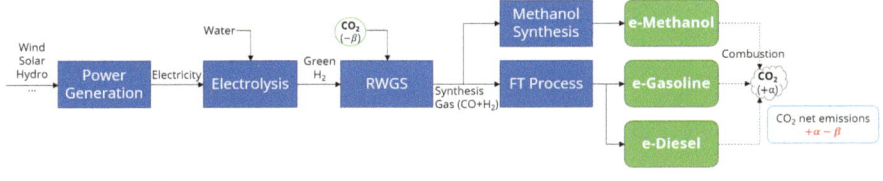

Fig. 6.6 Concept of e-fuel

that you can produce carbon monoxide from carbon dioxide and hydrogen.

$$WGS \text{ reaction: } CO + H_2O \rightarrow CO_2 + H_2$$
$$RWGS \text{ reaction: } CO_2 + H_2 \rightarrow CO + H_2O$$

Using these technologies, it is possible to synthesize alternative synthetic fuels using green hydrogen and carbon dioxide. This synthetic fuel produces carbon dioxide during combustion $(+\alpha)$. However, if no greenhouse gases are produced in the production process of the green hydrogen in the raw material, and the CO_2 in the raw material comes from the atmosphere $(-\beta)$, the net GHG emissions can be reduced to $(\alpha-\beta)$, just like biofuels. Ideally, if $\alpha \approx \beta$, the synthetic fuel can be considered as a carbon neutral fuel. This is the concept of e-fuels (electro fuel) as shown in Fig. 6.6.

E-fuels can be used in the same way as conventional petroleum fuels such as gasoline and diesel, which means that petroleum-based vehicles and facilities can be used to reduce GHG emissions without major changes to current facilities. For this reason, it is currently receiving a lot of attention, and there are many related research and projects.

6.8 Is E-fuel an Environmentally Friendly Fuel?

So, if we could just switch all of fuels to e-fuels as soon as possible, we would solve the problem of GHG emissions, so why don't we do it? Firstly, we need green hydrogen to produce e-fuels. Using synthetic fuels produced from fossil fuels instead of green hydrogen, is still the same as using fossil fuels, just in a different form, and therefore does not reduce GHG emissions. As mentioned in Sect. 5.3, green hydrogen is currently too expensive and there is a shortage of production. In other words, green hydrogen is a prerequisite for e-fuels, and if green hydrogen is not economically viable, e-fuels also will not be viable either.

Let's assume that green hydrogen becomes economically viable at some point in the future. Will that solve all the problems? No. In addition to green hydrogen, another raw material, the carbon dioxide, is needed to produce e-fuels. Depending on

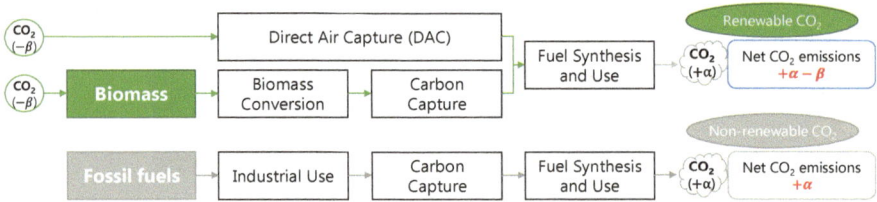

Fig. 6.7 Net CO_2 emissions of a synthetic fuel produced from renewable CO_2 and non-renewable CO_2

where the carbon dioxide in the raw material comes from, the WtW GHG emissions of e-fuels will change again.

There are two types of carbon dioxide available for the synthesis of e-fuels, as shown in Fig. 6.7. The first is non-renewable CO_2, which is the carbon dioxide captured from facilities that burn fossil fuels. The second is renewable CO_2, which is captured from sources that are not based on fossil fuels, such as biogas or direct air capture (DAC). When using fuels synthesized from renewable CO_2 based on non-fossil fuels, the net CO_2 emissions (α–β) can be reduced, because the amount of CO_2 emitted by combustion is added ($+\alpha$), but the amount of CO_2 absorbed from the atmosphere is subtracted ($-\beta$). For this reason, we can say that e-fuels have the GHG reduction effect as biofuels. However, if carbon dioxide is captured from an exhaust gas using fossil fuels such as coal, oil, and natural gas and used to make synthetic fuels, the use of the synthetic fuel is no different to the use of fossil fuels. For this reason, there is currently a debate on how to decide on the criteria for e-fuels.

There are currently no IMO criteria for e-fuels, and discussions are expected to continue. In EU, the FuelEU maritime has defined that renewable fuels of non-biological origin (RFNBO) should only be recognized as a renewable alternative fuel if they have at least 70% reduction in WtW GHG emissions compared to the fossil fuel baseline. If this criterion were to be applied, fuels synthesized from carbon dioxide captured from fossil fuels, even if they are synthesized with green hydrogen, would be unlikely to qualify as e-fuels.

Figure 6.8 shows examples of the WtW GHG intensity of methanol produced from green hydrogen and different CO_2 sources. For e-methanol produced from carbon dioxide captured from biomass or geothermal power plants, a 70%+ reduction of GHG intensity is possible. However, the WtW GHG emissions intensity of methanol, synthesized from carbon dioxide captured from a coal-fired power plant, is about 33 g_{CO_2eq}/MJ, which does not meet the 70% reduction requirement despite the use of green hydrogen.

While it is still difficult to say exactly what the international criteria will be set for e-fuels in the future and when they will be applied, it seems likely that the criteria will be progressively tightened toward 100% GHG reduction, as long as the IMO goal is to achieve net-zero emissions in the long term. It is therefore likely that, as with other fuels, the GHG intensity of e-fuel production and use will be subject to quantitative assessment.

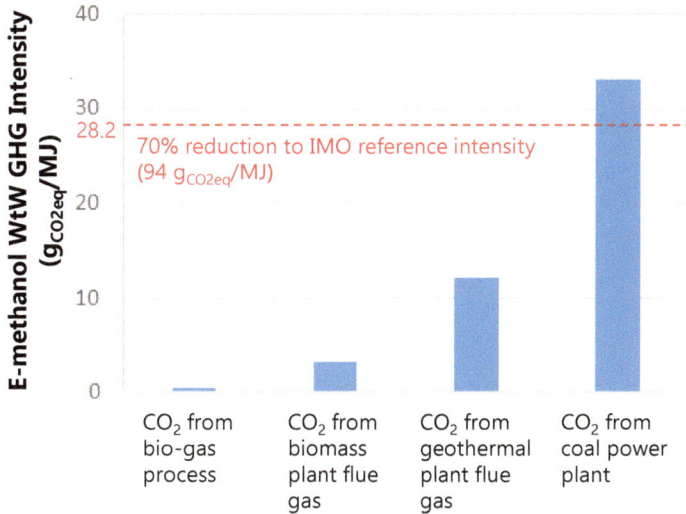

Fig. 6.8 WtW GHG emissions intensity of methanol synthesized from green hydrogen and various CO_2 sources [2]

References

1. Directive (EU) 2018/2001 of the European Parliament and of the Council of 11 December 2018 on the promotion of the use of energy from renewable sources. https://eur-lex.europa.eu/legal-content/EN/TXT/?uri=uriserv:OJ.L_.2018.328.01.0082.01.ENG
2. IRENA & Metanol Institute, *Innovation Outlook: Renewable Methanol* (International Renewable Energy Agency, 2021)
3. ISO 14067:2018, Greenhouse gases—carbon footprint of products—requirements and guidelines for quantification (2018). https://www.iso.org/standard/71206.html
4. B. Kampman, U.R. Fritsche, Better use of biomass for energy. IEA Bioenergy (2010)
5. MEPC.1/Circ.905, Interim guidance on the use of biofuels under Regulations 26, 27 and 28 of MARPOL ANNEX VI (DCS and CII) (2023)
6. RESOLUTION MEPC.376(80), Guidelines on life cycle GHG intensity of marine fuels (LCA guidelines) (2023)
7. E. Terazono, J. Jacobs, 'Diesel vs doughnuts': new biofuel refineries squeeze US food industry. Financial Times (2021). https://www.ft.com/content/b5839a04-a06a-49c1-8622-2974cbb9a84a

Chapter 7
Carbon Capture, Utilization, and Storage (CCUS)

Abstract Carbon capture, utilization, and storage (CCUS) is a combination of technologies for managing CO_2 emissions. CCUS includes technologies that capture CO_2 from sources of CO_2 emissions and either store it in sequestered locations or convert it into other useful materials to prevent the release of CO_2 into the atmosphere. CCUS has a long history, and the technologies to capture, transport, utilize and store CO_2 are relatively mature compared to other alternative fuels, so it can be applied to conventional systems without major modifications. Onboard carbon capture and storage (OCCS) is the application of CCUS technologies to a ship, to reduce the CO_2 emissions from a ship to the atmosphere by capturing and storing CO_2 onboard. OCCS also faces the challenges of a lack of regulation, high energy consumption, high costs, and insufficient infrastructure for offloading and transporting CO_2 to a final storage site. Nevertheless, to extend the lifetime of existing ships using fossil fuels and as a part of supply chains for sustainable alternative fuels, OCCS needs to be considered as a bridge technology to get through the transitional period until the era of fully sustainable alternative fuels arrives.

Keywords CCUS · CO_2 capture · CO_2 utilization · CO_2 storage · CO_2 transport · Onboard carbon capture and storage · OCCS

7.1 What Is CCUS?

Carbon capture, utilization and storage (CCUS) is a collective term for a group of technologies that capture CO_2 from large CO_2 sources and either store it in a sequestrated location, or convert it to another material to prevent its release into the atmosphere, as shown in Fig. 7.1. Initially, only technologies that capture and store CO_2 were referred to as carbon capture and storage (CCS), but the term CCUS is becoming more common as there is a growing consensus that technologies that store and utilize CO_2 have the same purpose.

Figure 7.2 shows CCUS facilities in operation by application from 1980 to 2021 [1]. While the amount of CO_2 captured before 2000 was small and applications were

© The Author(s), under exclusive license to Springer Nature Switzerland AG 2025   67
Y. Lim, *Alternative Fuels for Environmentally-Friendly Ships*,
SpringerBriefs in Applied Sciences and Technology,
https://doi.org/10.1007/978-3-031-85082-0_7

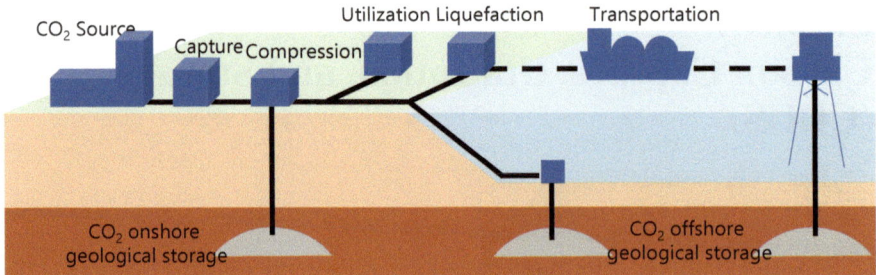

Fig. 7.1 Concept of CCUS

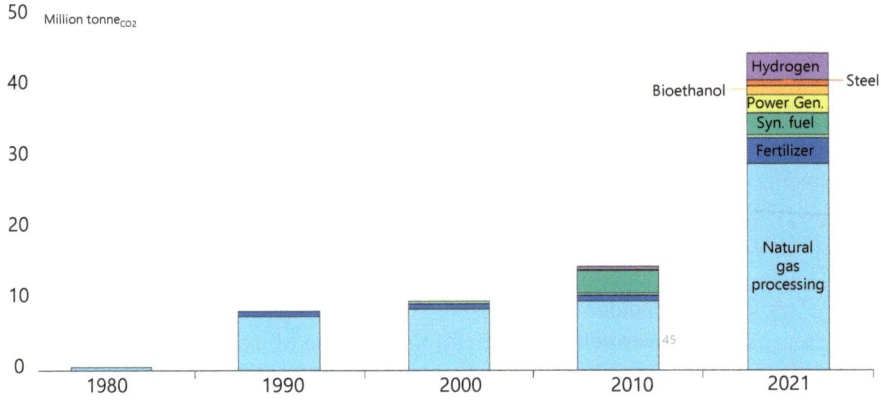

Fig. 7.2 CCUS facilities in operation by application, 1980–2021 [1]

mostly limited to natural gas processing, by the 2020s, annually more than 40 million tons of CO_2 were being captured from a variety of processes.

CCUS technology has a long history. The original purpose of CO_2 separation technology was to remove CO_2, not to capture it. As discussed in Chap. 4, natural gas present in a gas field within a geological formation contains hydrocarbons such as methane, as well as impurity gases such as nitrogen and carbon dioxide. High CO_2 content may degrade the quality of the natural gas and cause corrosion in pipelines, so they need to be separated. The process for removing acid gases, including carbon dioxide, during natural gas production is called an acid gas removal unit (AGRU) and has been in commercial use since the 1970s.

CO_2 injection into the geological formation has also been used commercially for many years. When producing crude oil from an oil field, the amount of crude oil that can be produced from natural reservoir pressure is typically around 10% of the total amount in the field. Various techniques have been developed to compensate for this pressure and one of them is to inject high-pressure CO_2 into fields to increase oil production, known as CO_2 Enhanced Oil Recovery (EOR), which has been used commercially since the 1980s.

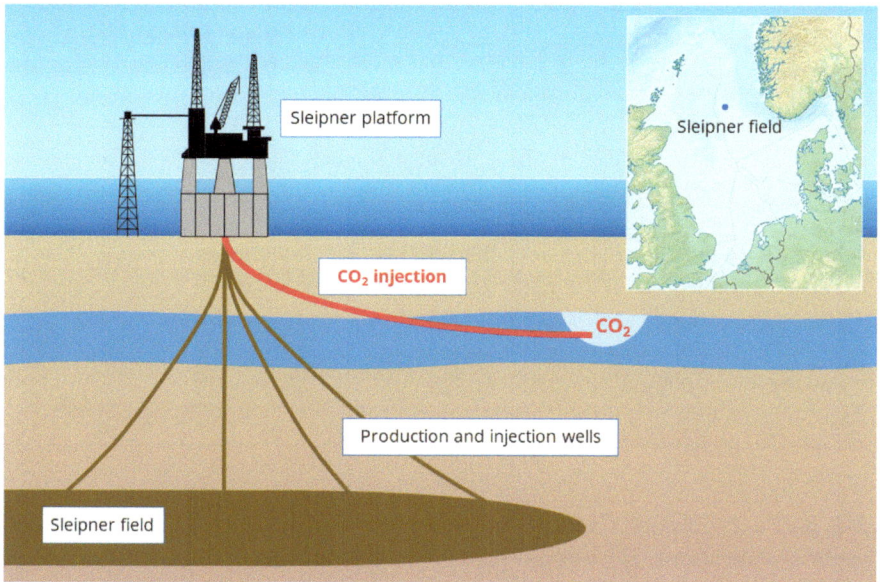

Fig. 7.3 Sleipner CCS project

One of the earliest commercial CCUS projects was the Sleipner project in Norway, which has been capturing and storing carbon dioxide since 1996. The EU began discussing a carbon tax in the early 1990s, but it could not achieve an EU-wide agreement. However, Norway, introduced an actual carbon tax in 1991, with industries paying a percentage tax based on their carbon dioxide emissions. The Sleipner gas field is an offshore gas field located in the North Sea between the UK and Norway, and although the gas reserves are large, the natural gas contains high portion of carbon dioxide. The Sleipner CCS project was planned to separate the carbon dioxide from the platform and inject it into the aquifer (a layer of groundwater containing salts and minerals) for storage as shown in Fig. 7.3. As a result of the success of this project, various CCUS projects are currently being planned and demonstrated.

7.2 CO$_2$ Capture

There are various carbon capture technologies, and the most common ones are presented here. In general, carbon capture technologies can be categorized as post-combustion capture, pre-combustion capture, and oxyfuel combustion, depending on the point of capture, as shown in Fig. 7.4.

Post-combustion capture is a technology that can be applied to the conventional combustion systems that use fossil fuels, such as engines or boilers. The capture systems are installed after the combustion systems to capture CO$_2$ from the exhaust

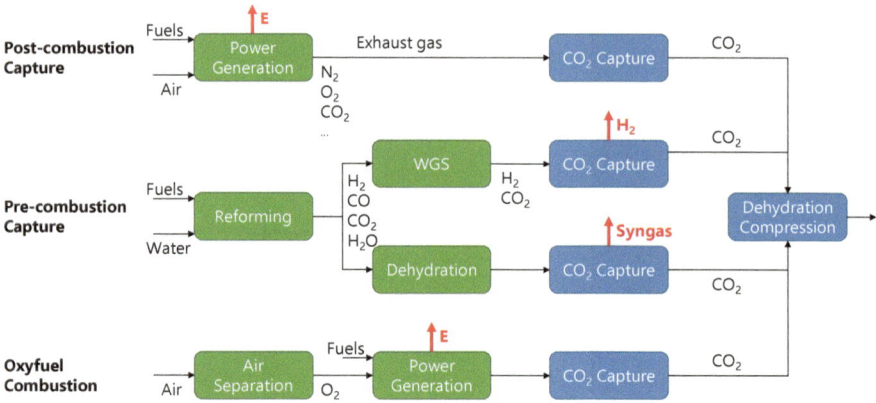

Fig. 7.4 CO$_2$ capture technologies

gas. It is easy to apply because existing fossil fuel engines can be used without major modifications. However, it requires additional energy consumption and pre-treatment of various impurities that may be contained in the exhaust gas.

Pre-combustion capture refers to the reforming of fossil fuels into other substances that can generate energy without emitting CO$_2$, and capturing the carbon dioxide generated in the process before the energy is used. A typical example of pre-combustion capture is the production of blue hydrogen, which involves reforming natural gas to produce hydrogen and capturing the carbon dioxide generated in the reforming process. It has the advantage of lower capture costs due to the relatively high concentration of CO$_2$ and low impurities in the reforming gas, but it requires a special process to reform the fuel prior to combustion (e.g., the steam methane reforming process for hydrogen production), which makes it difficult to apply to existing conventional combustion systems.

Oxyfuel combustion is a combustion system using only pure oxygen, rather than air. Ideally, if pure oxygen is supplied, complete combustion can be induced, generating only carbon dioxide and water. Water and carbon dioxide are easily separated into gas and liquid even at room temperature, so it is theoretically possible to obtain high-purity carbon dioxide without complex separation systems. However, this requires a specialized facility designed to ensure that the fuel is completely combusted with pure oxygen, making it difficult to apply to existing conventional combustion systems.

7.3   CO$_2$ Absorption

There are various technologies to capture CO$_2$, but the most commonly used method is the CO$_2$ absorption using materials that can physically or chemically absorb CO$_2$. For example, water absorbs CO$_2$ and becomes as sparkling water (carbonated water).

The water molecule reacts with CO_2 to form carbonic acid (H_2CO_3), which dissociates in aqueous solution into water in the form of bicarbonate ions (HCO_3^-) and carbonate ions (CO_3^{2-}) as shown in Fig. 7.5. However, the amount of carbon dioxide dissolved in pure water is very small, so it is more efficient to add other absorbents that can absorb carbon dioxide more effectively than water.

The most commonly used substances as CO_2 absorbents are amines, which are derivatives of ammonia (NH_3), wherein one or more hydrogen atoms are replaced by a substituent. When the substituent is an alcohol group, it is called an alkanolamine, such as Mono-Ethanol-Amine (MEA), Di-Ethanol-Amine (DEA), and Methyl-DiEthanol-Amine (MDEA) as shown in Fig. 7.6.

Amines have a high affinity for CO_2 at low temperatures, so they react with CO_2 to form water-soluble ions, allowing CO_2 to be absorbed into the aqueous amine solution as shown in Fig. 7.7. At higher temperatures, the reverse reaction occurs, releasing CO_2 as a gas. Amine-based solutions can therefore be used as efficient CO_2 absorbents.

Figure 7.8 illustrates the basic principle of a CO_2 absorption process using an absorbent solution. The CO_2 absorption process typically consists of two columns, the first of which is an absorber (absorption column) that sprays an absorbent solution from the top. Exhaust gas containing CO_2 is injected at the bottom and through the

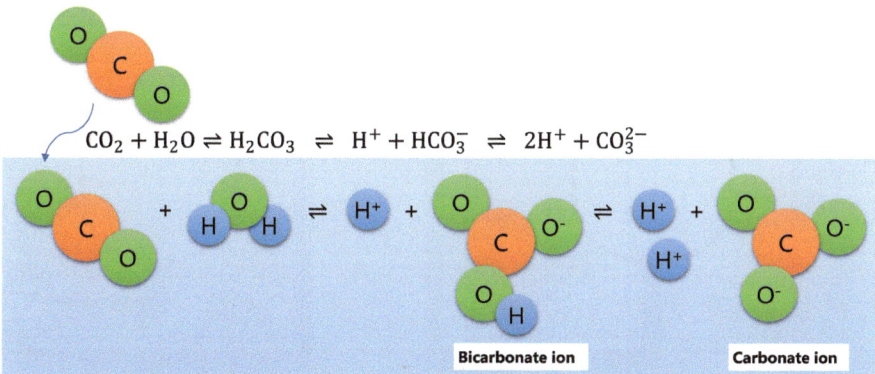

$$CO_2 + H_2O \rightleftharpoons H_2CO_3 \rightleftharpoons H^+ + HCO_3^- \rightleftharpoons 2H^+ + CO_3^{2-}$$

Fig. 7.5 CO_2 absorption into water

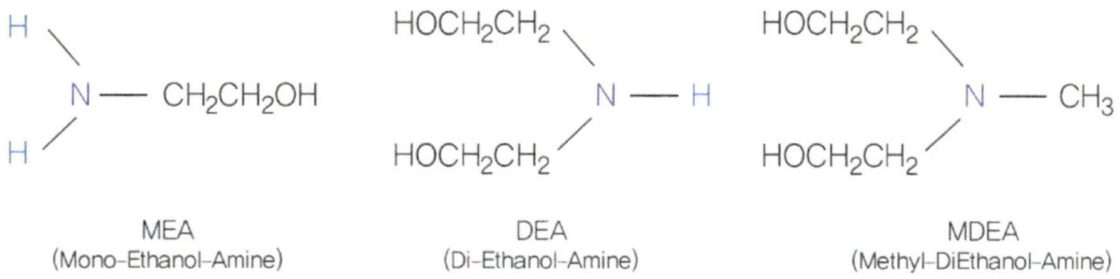

MEA
(Mono-Ethanol-Amine)

DEA
(Di-Ethanol-Amine)

MDEA
(Methyl-DiEthanol-Amine)

Fig. 7.6 Representative amine compounds

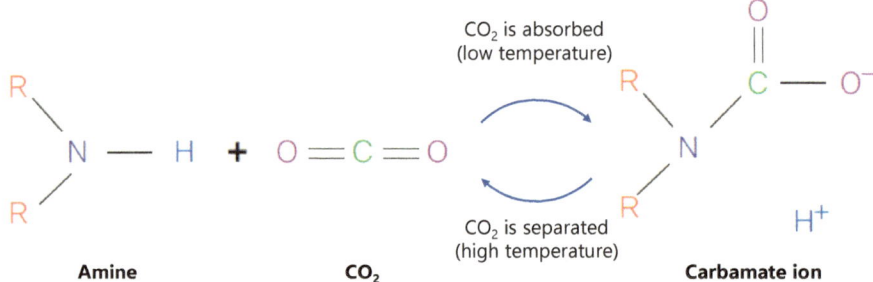

Fig. 7.7 Example of CO_2 absorption mechanism of amine

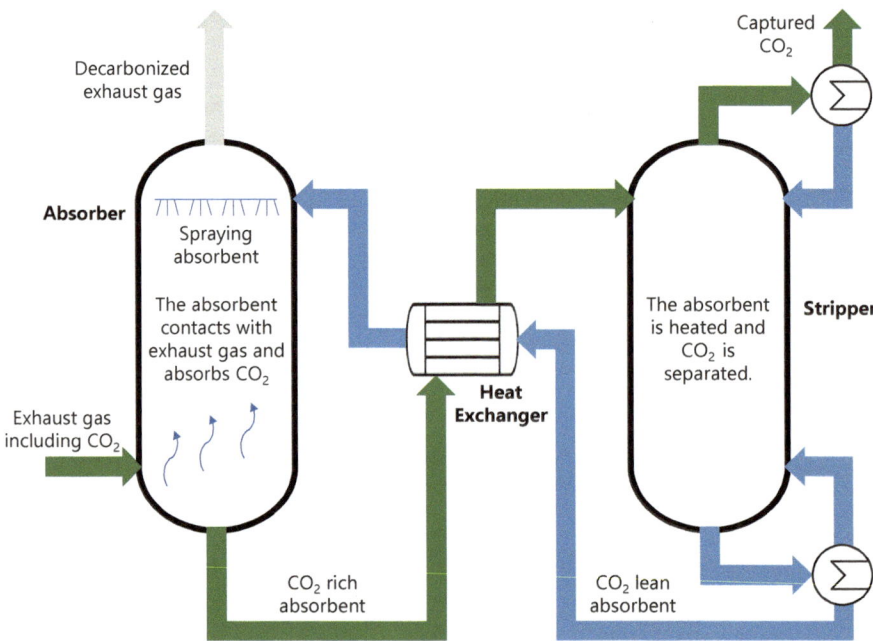

Fig. 7.8 Flow diagram of a CO_2 absorbing process

internal structure, the absorbent and the exhaust gas come into contact, and the CO_2 in the exhaust gas is absorbed by the absorbent. The CO_2-removed exhaust gas is discharged from the top of the column, and the CO_2-rich absorbent solution flows out of the bottom of the column. After heat exchange, then the CO_2-rich absorbent solution is sent to the stripper (regeneration column) and a reverse reaction occurs to separate CO_2 at high temperatures by heating in the reboiler. High purity CO_2 is captured from the top of the stripper and the regenerated CO_2-lean absorbent solution is discharged from the bottom of the stripper. The CO_2-lean absorbent solution is then recycled to the absorption tower via heat exchange and cooling. In this way, it is possible to continuously capture carbon dioxide from the exhaust gas.

The CO$_2$ absorption process has a long history and has been commercially demonstrated by various institutes, so its performance can be guaranteed. However, a large amount of heat is required to regenerate the absorbents, and if the energy consumed for this heat comes from fossil fuels, carbon dioxide is generated again. A lot of research is underway to develop better absorbents that require less energy to regenerate, and to reduce the amount of heat consumed by recovering waste heat from exhaust gases.

7.4   CO$_2$ Transport

CO$_2$ is a widely used substance in many industries, and because of this demand, the transport of CO$_2$ is an industrially mature technology that has been transported thousands of kilometers around the world. For example, we can easily see dry ice (solid carbon dioxide) all around us. Another industrial example of the need for large amount of CO$_2$ is CO$_2$ Enhanced Oil Recovery (CO$_2$ EOR), a technique for increasing crude oil production by injecting CO$_2$ at high pressure into oil fields. Onshore, carbon dioxide is mainly transported via pipeline as a high-pressure gas compressed to more than 100–200 bar. For safety reasons, pipeline transport is usually subject to impurity removal requirements, as shown in Table 7.1.

Liquefied carbon dioxide (LCO$_2$) carriers can be used where there is no land connection, or where long-distance transport is required. However, existing LCO$_2$ carriers are small-scale vessels with a capacity of ~ 1000 tonnes, and there were no commercially operated large LCO$_2$ carriers for CCUS. In 2023, HD KSOE contracted for two 22,000 m^3 LCO$_2$ carriers, and reported that HD Hyundai Mipo will build and deliver from the second half of 2025.

LCO$_2$ has thermodynamic properties that are different from other substances. Substances exist in solid, liquid and gas phases and the temperature and pressure at which each phase exists is unique to the substance. The phase change from solid to gas or directly from gas to solid without passing through a liquid at normal pressure is called sublimation, and carbon dioxide is a typical substance that sublimates at normal pressure.

Table 7.1   CO$_2$ specifications example for pipeline transport

Item	Typical specification	Remark
CO$_2$ purity	95+%	
Hydrocarbons	< 5%	
Oxygen, sulfur compounds	< 10–100 ppm	Causes corrosion. Sulfur compounds are toxic
Water	< 10–20 ppm	Causes corrosion and hydrate formation

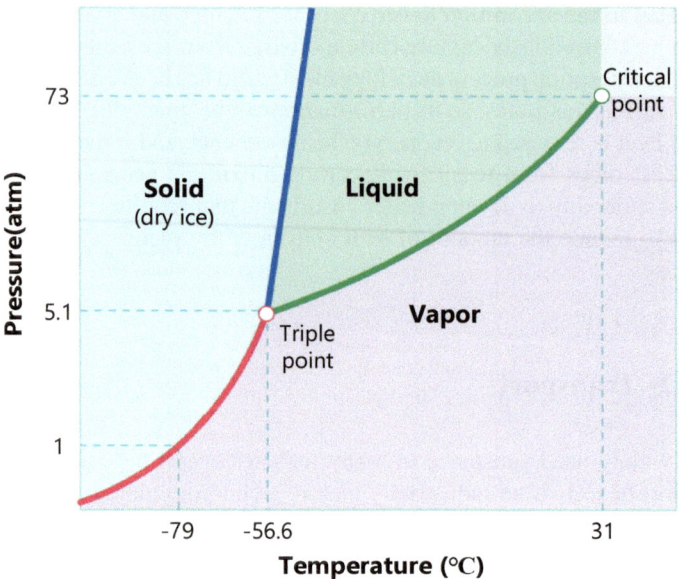

Fig. 7.9 PT diagram of CO_2

CO_2 PT diagram shows the phase change of CO_2 with temperature and pressure, as shown in Fig. 7.9. In the case of carbon dioxide, it has been experimentally confirmed that the triple point at which solid–liquid–gas can coexist is about 5.1 atm and − 56.6 °C. This means that a pressure of at least 5.1 atm is required to liquefy carbon dioxide. For this reason, small LCO_2 carriers currently in use store LCO_2 at typically 10–20 atm. However, the higher the pressure of the storage vessel, the more expensive it is, so the large-scale LCO_2 carriers are considering a low pressure storage system within 7–10 bar. There are also opinions that it could be stored as dry ice at normal pressure rather than increasing the pressure. However, unlike liquids and gases, which can be continuously loaded and unloaded through pipes, solid materials are not easy to load and unload continuously in large quantities, so it would be difficult to apply to large-scale LCO_2 carriers.

7.5 CO_2 Storage

Even if CO_2 is captured, it would be pointless to release it back into the atmosphere, so CCUS technologies ultimately need a means of keeping it out of the atmosphere or converting it into other substances. The most widely recognized method is underground geological storage, where CO_2 is injected and stored in geological formations that have the geological structure to keep it out of the atmosphere.

Examples of structures that can store CO_2 include depleted oil/gas fields, or saline aquifers, which are layers of groundwater containing salt and impurities that make it unsuitable for drinking and cooking. The many CCUS projects currently in operation use enhanced oil recovery (EOR) to store carbon dioxide, which injects CO_2 into the formation to increase crude oil production. While this is economically feasible, it is relatively ineffective at sequestering CO_2 because some of the injected CO_2 may be released with the crude oil being produced.

CO_2 injected into a formation is trapped in the formation by the following processes. The first step is structural trapping, where the injected CO_2 collects in the lower part of the cap rock, which has a low permeability and is therefore difficult to penetrate. The second step is residual trapping, where the CO_2 is absorbed by capillary action in the pores of the rock in the formation. The third step is solubility trapping, where the CO_2 is dissolved in groundwater. Finally, CO_2 reacts with mineral ions (calcium, potassium, etc.) in the formation forming carbonates (such as calcium carbonate), and are mineralized and trapped in the formation. This is the same principle by which calcium carbonate in groundwater precipitates, forming stalactites in limestone caves.

Another issue with storing CO_2 in formations is the safety concern of whether the CO_2 injected into the formation would leak out and cause another accident. In 2017, Norwegian Oil Company Equinor (formerly Statoil) published the results of 20 years of monitoring of the CO_2 injected into the saline aquifer at the Sleipner CCS project since 1996, showing that the CO_2 is immobilized in the formation as expected, with no leaks. However, the possibility of an unexpected accident still remains, so safety precautions are essential.

7.6 Northern Lights Project

There are some developed sites for geological storage of CO_2, but they are allocated to specific companies or countries. For countries or companies that do not have suitable sites for large-scale CO_2 storage, it is more difficult to implement CCS projects. To overcome this problem, there are currently a few cross-border CO_2 storage projects planning international CO_2 transport and storage. One such project is the Northern Lights project in Norway.

The Northern Lights project aims to build a CO_2 hub terminal and inject CO_2 into the offshore subsea geological formations in the North Sea as shown in Fig. 7.10. The captured CO_2 will be transported by LCO_2 carriers, and the onshore hub terminal will store the transported LCO_2, pump it to high pressure, and send it through approximately 100 km of subsea pipeline to the subsea injection well. Finally, the CO_2 will be injected into subsea reservoirs located at a depth of ~ 2.6 km below the seabed surface.

In phase 1, CO_2 will be transported from onshore sources, such as cement plants in southern Norway, and injected. Once the demonstration is complete, the plan is to receive and inject CO_2 from neighboring countries in Europe as shown in Fig. 7.11.

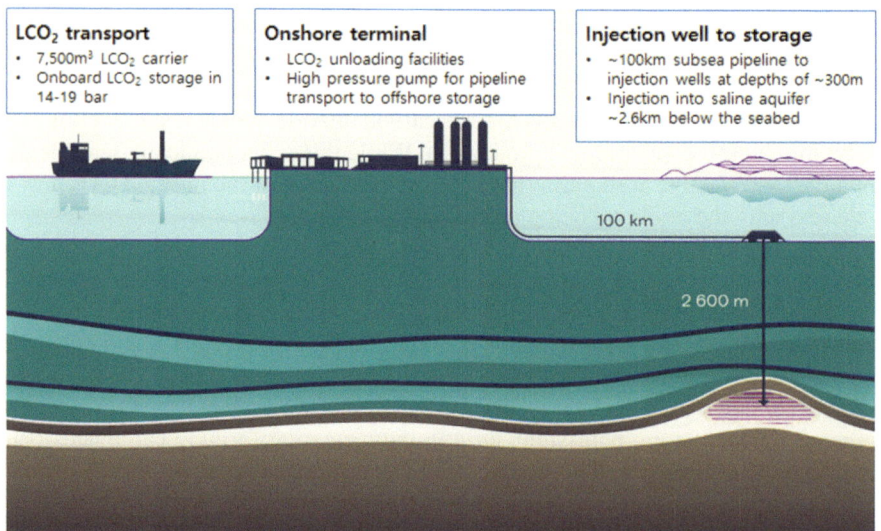

Fig. 7.10 Concept of the Northern lights project [5]

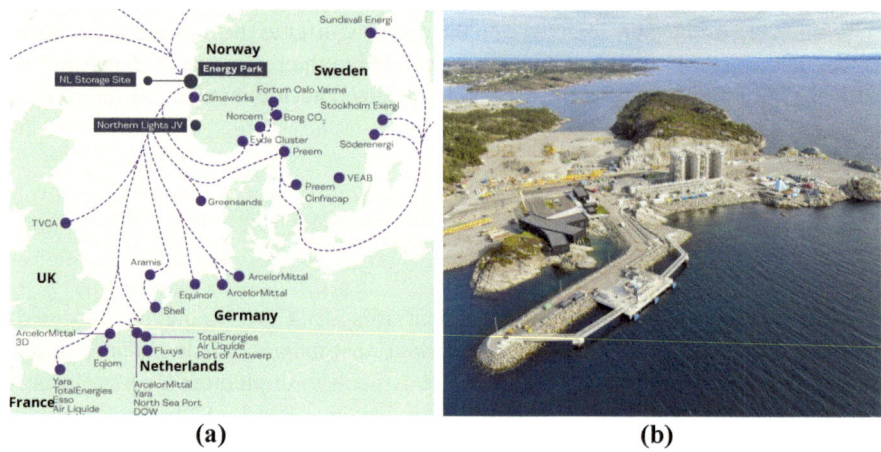

(a)   **(b)**

Fig. 7.11 a Potential CO_2 source near the Northern Lights site and **b** LCO_2 terminal under construction [3–5]

In 2024, construction of the LCO_2 hub terminal is expected to be completed and the first phase of demonstration injection will begin. This project is attracting a lot of attention because, if it is successful, it could open up the possibility of cross-border CO_2 trading projects in countries in earnest.

7.7 Economic Feasibility of CCUS and DAC

While there are technical challenges to making CCUS work in practice, there are also major economic challenges. CO_2 capture requires a lot of energy, which makes it expensive. There are many factors that affect the cost, the most critical one is the concentration of CO_2 in the capture source; the higher the concentration of CO_2 in the capture source, the cheaper it is to capture. Currently, the CO_2 capture cost is estimated to be around 20–150 $\left(\text{USD/tonne}_{CO_2}\right)$ or more. The cost of CO_2 transport is estimated at 5–30 $\left(\text{USD/tonne}_{CO_2}\right)$ per ton depending on the distance, and the storage cost is estimated at 10–50 $\left(\text{USD/tonne}_{CO_2}\right)$ depending on the depth of the site. Voluntarily incurring these high costs to capture and store CO_2 is a challenge for companies that need to make a profit to survive. This is why there is an ongoing debate about regulating CO_2 emissions, introducing a carbon tax and a system of trading carbon credits.

In Sect. 6.8 it was discussed that e-fuels based on CO_2 captured from biomass, or Direct Air Capture (DAC), will qualify as an environmentally friendly fuel in the future. DAC refers to technologies that capture CO_2 directly from the air, and is currently technically feasible by using absorption processes or adsorption processes. If this is possible, why not simply use DAC on a large scale to capture carbon dioxide from the atmosphere without going through the complex processes? This is also a techno-economic question. Figure 7.12 shows the CO_2 capture cost by source, which is estimated by the International Energy Agency (IEA). It can be seen that the cost of CO_2 capture from sources with relatively high CO_2 concentrations is estimated to be relatively low, below 50 $\left(\text{USD/tonne}_{CO_2}\right)$, whereas the DAC estimates very high costs of over 200–250 $\left(\text{USD/tonne}_{CO_2}\right)$. This is because the concentration of atmospheric carbon dioxide is so low, in the order of a few hundred parts per million, that it requires a lot of energy to capture it with high purity.

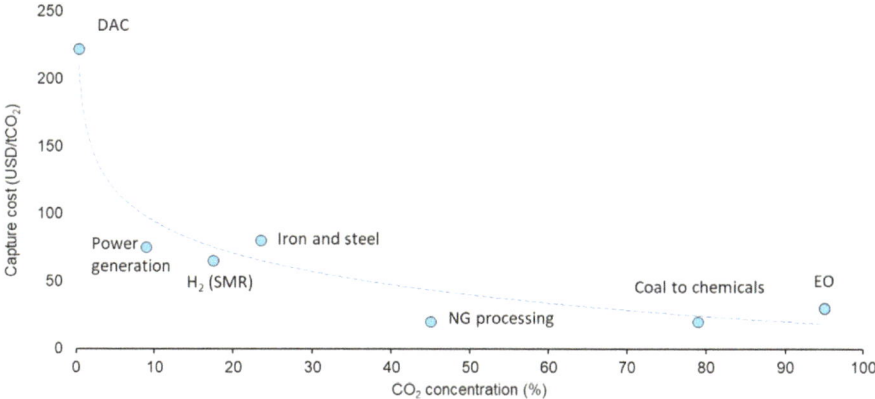

Fig. 7.12 CO_2 capture cost at varying CO_2 concentrations [2]

7.8   CO$_2$ Utilization

A number of technologies have been developed to produce a variety of materials from CO$_2$. As described in Sect. 6.7, e-fuels can be produced from CO$_2$, and many other industrial chemicals and building materials can be produced from CO$_2$, as shown in Fig. 7.13.

So why are we discussing CO$_2$ capture and storage in the formation rather than simply turning it into a useful product? There are two main reasons. First, CO$_2$ is a thermodynamically stable substance, and most conversion processes that change it into other substances require energy, to make high temperature and pressure. If the energy comes from fossil fuels, the conversion of CO$_2$ into other substances will again generate GHGs. If the amount of GHGs generated is greater than the amount of carbon dioxide consumed in the conversion, the conversion technology is not reducing GHGs, but rather emitting them. Table 7.2 is an example of a study that estimates the GHG emissions per unit mass of material produced by different CO$_2$ utilization technologies. It can be seen that CO$_2$ conversion by using renewable energy, such as solar, results in a reduction in GHG emissions, but using electricity or hydrogen produced from fossil fuels results in an increase in GHG emissions. In other words, it is not the CO$_2$ conversion itself that is important, but whether the entire conversion process results in a net reduction in GHG emissions when all energy consumption is considered.

Second, there is the issue of economics. Figure 7.14 shows the results of a study that estimates the ratio of the estimated production cost of substances using CO$_2$ utilization technologies to the market selling price. In most cases, the ratio is greater

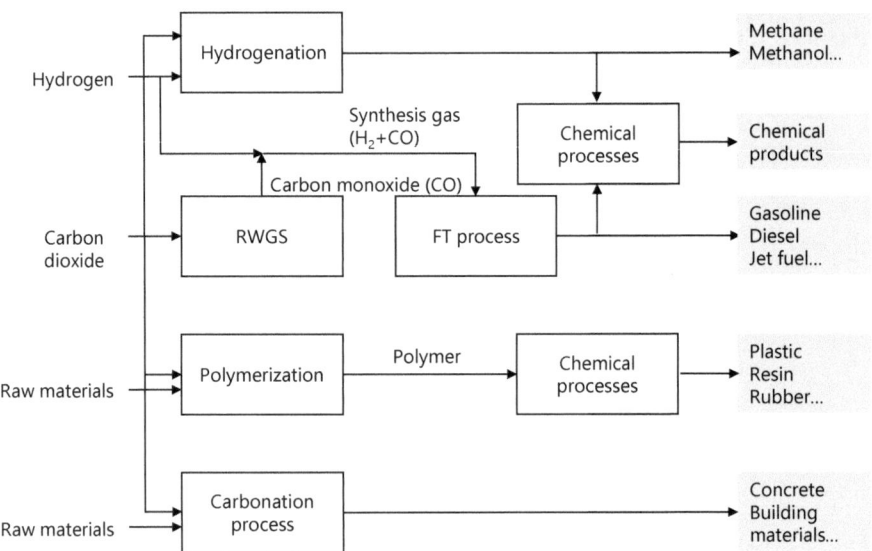

Fig. 7.13 Examples of various products from CO$_2$

Table 7.2 Net GHG emissions per unit mass of various products

Product	Net GHG emissions per unit mass of the produce $(t_{CO_2eq}/t_{Product})$	Remark
Formic acid	− 0.31 (reduction)	Using solar power
	+ 3.42 (increase)	Using grid power (based on fossil fuels)
Methanol	− 0.87 (reduction)	Using hydrogen, from renewable energy
	+ 3.91 (increase)	Using hydrogen, from fossil fuels
Methane	− 2.44 (reduction)	Using hydrogen, from renewable energy
	+ 9.30 (increase)	Using hydrogen, from fossil fuels

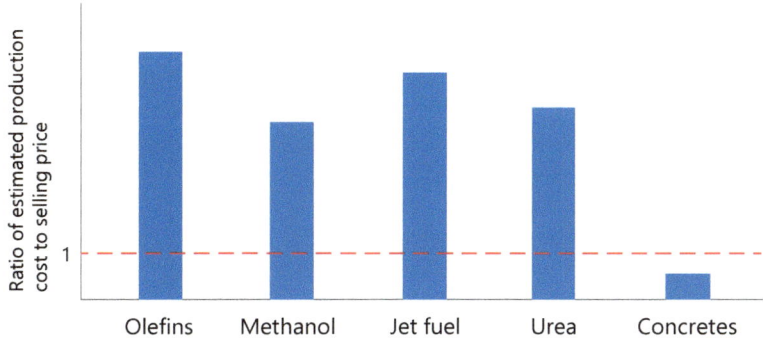

Fig. 7.14 Ratio of estimated production cost to selling price for various products [6]

than 1, which means that it is more expensive to produce it than the market value, and therefore not commercially viable. In other words, many CO_2 conversion technologies are not yet commercially viable without additional financial support. For these reasons, although CO_2 utilization technologies have great potential as a solution to GHG mitigation, but much research is still needed.

7.9   Onboard Carbon Capture and Storage (OCCS)

Then, is it possible to use CCUS technologies on a ship to reduce CO_2 emissions from a ship? This is the motivation of the introduction of Onboard Carbon Capture and Storage (OCCS). Here, the "storage" means onboard temporary storage of CO_2, not permanent storage in geological formations. Technologies such as carbon capture or

CO_2 liquefaction have been used commercially for a long time, so it's not unreasonable to apply them to ships. In 2021, the Japanese shipping company Kawasaki Kisen announced that it had installed and tested carbon capture systems on a ship, demonstrating that onboard carbon capture is possible. More recently, OGCI, TNO, Wärtsilä, HD Hyundai, Samsung Heavy Industries and Hanwha Ocean are also developing and demonstrating onboard carbon capture systems (OCCS). Figure 7.15 shows that an example of OCCS systems.

Unfortunately, currently OCCS systems are not yet commercially viable. There are three main issues that need to be addressed. First, international regulations for OCCS need to be established. Currently, there are no international regulations at the IMO on how to quantify the reduction of GHG emissions by OCCS systems. At the 81st MEPC in March 2024, it was agreed to establish a correspondence group to review the reduction potential and additional energy consumption of OCCS systems and to develop certification guidelines, so there may be a way forward in the near future.

Second, a supply chain needs to be established to store or utilize the CO_2 captured from ships. Captured CO_2 can't be stored on a ship forever, and if it is released back into the atmosphere, the CO_2 capture becomes meaningless. Therefore, the captured CO_2 should be able to be offloaded somewhere, and the offloaded CO_2 needs to be transported to a site for a geological storage or utilization. At present, there are not enough of these sites, but if the recent CO_2 offloading and injection projects, such as the Northern Lights project described in Sect. 7.6 work well, this problem will hopefully be solved in the future.

Third, there is the economic issue; CO_2 capture and storage requires energy. All energy used on a ship should be generated within the ship, and currently fossil fuels such as HFO or MDO/MGO are used to generate the energy on a ship. Therefore, the systems installed to capture and liquefy CO_2, indirectly generate CO_2 again by consuming fossil fuels. Even if the waste heat in the exhaust gas is recovered as much

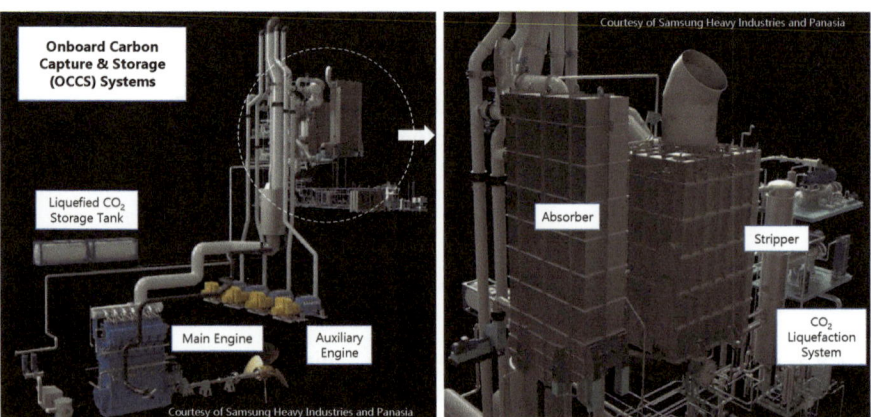

Fig. 7.15 Example of onboard carbon capture and storage systems

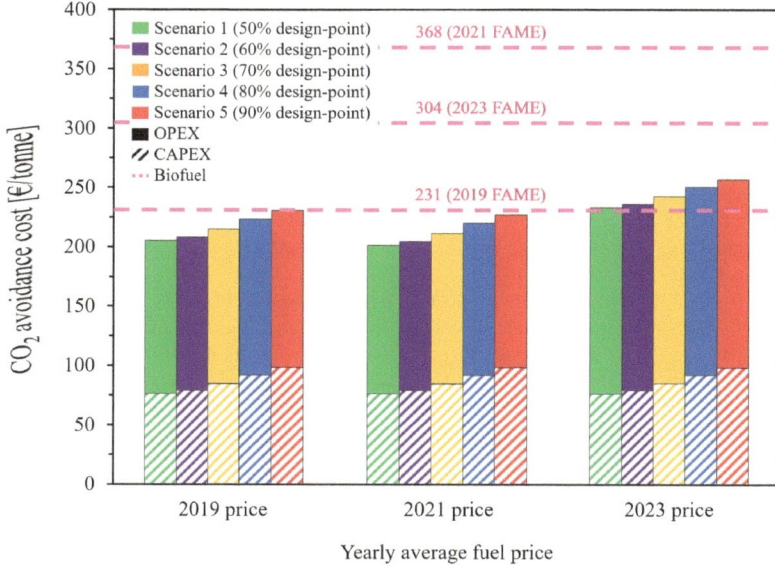

Fig. 7.16 Example of the research results on CO_2 avoidance cost for OCCS [7]

as possible, many studies have reported that high energy and costs are required to avoid GHG emissions from a ship, as shown in Fig. 7.16. In addition, the captured CO_2 has to be stored onboard until it can be offloaded, which reduces cargo capacity and results in additional economic losses. The economic feasibility of OCCS systems needs to be evaluated by considering these factors.

Despite these challenges, why is there still so much discussion about OCCS? First, there is no absolute solution for ships to reduce WtW GHG emissions. Although there is a lot of interest in alternative fuels, there are currently not enough viable alternative fuels for ships now, as discussed in Chaps. 5 and 6. and Second, there are many existing ships in operation using fossil fuels, and not all of them can be converted to use alternative fuels in the short term. If appropriate international regulations are introduced, the use of OCCS could extend the life of existing ships. Third, OCCS may be needed as part of a sustainable alternative fuel supply chain. For example, blue hydrogen is produced from natural gas, so requires CCUS as described in Chap. 5. If there is no CO_2 storage or utilization site close to the capture site, the captured CO_2 will need to be transported over long distances using CO_2 carriers. If this CO_2 carrier emits GHGs during transport, it may result in an increase of the GHG emission intensity of blue hydrogen from a WtW GHG emission perspective. This means that CO_2 carriers for sustainable alternative fuels such as blue hydrogen or blue ammonia are more likely to require OCCS systems. For CO_2 carriers, it is also relatively easy to store the captured CO_2 onboard because they already have CO_2 storage tanks. These are the reasons why discussions and research on OCCS systems continue.

References

1. IEA, About CCUS (2021). https://www.iea.org/reports/about-ccus. Accessed 30 Oct 2024
2. IEA, *Direct Air Capture: A Key Technology for Net Zero* (2022)
3. Northern Lights, Annual Report 2021 (2021). http://norlights.com/what-we-do/reports/. Accessed 30 Oct 2024
4. Northern Lights, Annual Report 2021 (2022). http://norlights.com/what-we-do/reports/. Accessed 30 Oct 2024
5. Northern Lights, Annual Report 2023 (2023). http://norlights.com/what-we-do/reports/. Accessed 30 Oct 2024
6. Bhardwaj et al. Opportunities and limits of CO_2 recycling in a circular carbon economy (2021)
7. J. Oh, D. Kim, S. Roussanaly, R. Anantharaman, Y. Lim. Optimal capacity design of amine-based onboard CO_2 capture systems under variable marine engine loads. Chem. Eng. J. **483**, 149136 (2024)